Global
Dumping
Ground

Global Dumping Ground

The International Traffic in Hazardous Waste

Center for Investigative Reporting
and
Bill Moyers

Seven Locks Press
Washington

Library of Congress Cataloging-in-Publication Data
Moyers, Bill D.
    Global dumping ground: the international traffic in hazardous
waste/ Center for Investigative Reporting and Bill Moyers.
    p. cm.
    Includes bibliographical references and index.
    ISBN 0-932020-95-X : $11.95
    1. Hazardous waste management industry. 2. Hazardous waste
management industry—Corrupt practices. I. Center for Investigative
Reporting (U.S.). II. Title.
HD9975.A2M69   1990
363.17—dc20   90-43079
                                    CIP

Manufactured in the United States of America
Cover illustration by Nancy Casalina, Video Arts, Inc.,
        San Francisco, CA
Typesetting by Edington-Rand, Inc.
Printed by McNaughton & Gunn, Inc., Saline, MI

Seven Locks Press is a Washington-based book publisher of non-fiction
works on social, political and cultural issues. It takes its name from a
series of lift locks on the Chesapeake and Ohio Canal.

For more information or a catalog:
    Seven Locks Press
    P.O. Box 27
    Cabin John, MD 20818
    (301) 320-2130

CONTENTS

Preface

The images refuse to go away: the distressed mother along the Mexican border, her children bathing in polluted waters; the Brazilian workers suffering the effects of lead poisoning; children in Taiwan wearing protective masks at school to guard against the fumes next door.

These are only a few of the human costs exacted by the world's new commerce in hazardous waste.

A disturbing, illicit side to this growing trade evokes other images as well: ghost ships laden with toxic cargo, midnight haulers, and shadowy characters in a business with few rules.

For four years, reporters and researchers at the Center for Investigative Reporting have tracked our collective descent into a global dumping ground. They have charted how industrial wastes are surreptitiously shipped abroad to the often unsuspecting and unprepared peoples of the Third World—to Panama, Zimbabwe, China, and other nations around the globe. As the following pages reveal, the effects can be devastating. Simply put, the world's wealthiest nations are exporting vast quantities of toxic waste, potential Love Canals and Minamatas for others to handle. One more burden for the poor of the world.

I became involved in this story in my role as a member of the Center's board of advisors. At the time, the Center had launched a documentary project for public television, entitled "Global Dumping Ground," and asked for my support. After examining their research, I knew they were onto something significant. The opportunity to join in was not to be missed. As Nat Hentoff recently remarked, reporting is the highest form of journalism, and the Center was setting forth to prove once again that investigative reporting can show us how the world truly works.

In the great tradition of American muckraking, the Center's reporters have adhered to the cardinal rule of the craft: follow the money. The motivation for the multinational corporations, the fly-by-night waste traders, and the assorted con artists involved in dumping is not to harm the environment, but simply to make a profit. As our documentary and this companion book make clear, the current system of environmental laws actually provides incentive for them to do so.

Here in the United States, for example, our laws in effect have encouraged industry to produce more waste, not less. Not long ago, some of the most lethal by-products of our industrial age were dumped into vast pits, forgotten until the people in nearby communities began to ask why their neighbors were so ill, why their children were dying. Now, finally, as the industrial world's environmental laws are being strengthened, such mindless dumping is increasingly a thing of the past. Our new environmental consciousness, alas, seems to end at each nation's borders. In the United States, as in most of the world, no laws stop those who would ship our hazardous wastes abroad to an uncertain disposal. Indeed, the entrepreneurs who offer to take away these poisons relieve their clients of a cumbersome, expensive obligation under American law. Once overseas, any remnant of responsibility, legal or moral, seems to vanish— chilling testimony to the old adage, "out of sight, out of mind."

What is so striking about the global trade in dumping is not only its extraordinary reach, but how unnecessary it all is. The vast mountains of hazardous waste need not be an inevitable by-product of how we live. As the chapters that follow will attest, it is not simply our way of life that produces such waste—it is our way of thinking. There are alternatives, and they do not require returning to the Pleistocene Age.

In this time of glasnost, as the threat of nuclear armageddon at last recedes, our governments are beginning to understand how quickly we are despoiling the earth, and how our pollutants respect no borders. Before the public can be enlisted in putting an end to this dirty business, however, the facts must be hauled from the shadows and stacked in plain view for all to see. In this book, the journalists at the Center for Investigative Reporting have performed a remarkable public service. Here are the facts, crying out for attention—and action.

Bill Moyers

Acknowledgments

This book is the result of a team effort on the part of everyone at the Center for Investigative Reporting. A project of this magnitude could not have been realized without many people making significant contributions.

As a companion to the public television documentary "Global Dumping Ground," this book is based largely on the extensive reporting done by Center reporters for that program. The impetus to develop the Center's first independent television documentary grew out of the response to our earlier work on waste exports by David Weir and Andrew Porterfield. Their efforts resulted in a series of reports in *The Nation*, on CBS News, and in publications worldwide, helping to spark international interest in the issue.

None of the Center's reporting on global dumping would have occurred, however, had it not been for the pioneering work of David Weir and Mark Schapiro in CIR's book, *Circle of Poison* (1981), about the dumping of banned and restricted pesticides overseas.

Major credit goes to CIR program director Sharon Tiller for having launched the documentary, refusing to give up when there were doubts, and raising nearly $500,000 for the program. Thanks also to CIR editor David Kaplan who crafted the original proposal to the Corporation for Public Broadcasting.

None of this would have been possible without the generosity of the following funders: The Corporation for Public Broadcasting, Public Broadcasting Service, The John D. and

Catherine T. MacArthur Foundation, The Florence and John Schumann Foundation, The George Gund Foundation, Threshold Foundation, Columbia Foundation, Wallace Alexander Gerbode Foundation, United Nations Environment Programme, The Tides Foundation/Pohaku Fund, Alida Rockefeller, BridgeBuilders Foundation, The Strong Foundation, Harle Montgomery, and Jonathan Z. Larsen. Nearly twenty other individuals also made donations to the program, and we thank them as well.

Josh Darsa and Jennifer Lawson at the Corporation for Public Broadcasting provided crucial encouragement for the documentary, and Bob Richter supplied advice when the Center knew little about the ways of independent production for public television. Gene Aleinikoff contributed legal counsel based on his years of experience in public broadcasting. Ed Davis and Judy Alexander at Pillsbury, Madison & Sutro gave their careful review to the manuscript and the documentary.

At FRONTLINE, David Fanning provided encouragement and key support in bringing "Global Dumping Ground" to a national audience. Thanks also go to James Morris and the staff at Seven Locks Press. They moved in record time to bring this book to life.

We appreciate the help of Jim Vallette and the many staff members of Greenpeace for their expertise on toxic-export issues. They were generous with their research and particularly valuable in helping to focus our research early on. We also want to thank Dr. Martin Abraham and sources at the International Organization of Consumers Unions (IOCU) in Penang, Malaysia, who provided us with many valuable articles and contacts on the global trade in hazardous products, technologies and waste; and Louis Blumberg, co-author of *War on Waste*, for his assistance.

Special thanks also go to the following individuals and organizations for their assistance: Tony Baldo, Bay Area

Video Coalition, Glenn and Diane Bergman, Andreas Bernstorff, William Brown, Robert Buechler, Brad Bunnin, Fire Chief Henry Campbell, William Carter, Charles Chen and Jackson Chen with the Coordination Council for North American Affairs, Joseph Chen, Jeff Chester, Dr. Eugene Chien, Charles and Jack Colbert, Robbie Conal, Dave Cook, James R. DeVita, Maria Erana, Lance Erickson, Ernest Esztergar, Michael Fellner, Marcy Fenton, Jock Ferguson and the *Globe and Mail* (Toronto), Bruce Gellerman, Dianne Gresser, Wendy Grieder, Yolanda Grizzard, Javier Gutierrez, Joel Hirschhorn, Leslie and John Hoge, Steve Hronek, Shielin Wong Hua, Marni Inskip, Ralph King, Jules Kroll, Bill Lambrecht, Larry Larson and family, Tom Layton, Gerald Lenoir, Richard Liebner, Li-Teh Lu, Rebecca Rosen Lum, Hugo Martinez, Father Jerry Martinson and the Kuangchi Program Service in Taiwan, Mike and Laura Michaelson, Jeff Millefoglie, Myra Ming, Monica Moore, Victor Navasky, Alice and Jules Perlmutter, Margaret Phillips, Drummond Pike, Liza Cohen Pike, Annie Pong, Michael Rabinowitz, John S. Redden, Padraic D. Riley and the Gannett Westchester Newspapers in White Plains, N.Y., Rosa Devora Rezo and family, Stanley Robertson, Miriam Rogow, Debbie Ross, Allen P. Rossi, Matt Rothman, Jack Rust, Kim Salyer, Maurilio Sanchez and family, Roberto Sanchez, Mark Schapiro, C. Shen, Susan Silk, Sandy Spencer, Laura Surtes, Michael Schwarz, Congressman Michael Synar and his staff, Sintesis Television, Tijuana, Mexico, Joan and Ron Turner, Liz Turner, Laurie Udesky, Steve Vale, Fanny Wang and Wayne Lo with the Thai Ping Metal Industrial Company, Dr. Jung-Der Wang, Rob Waters, Doug Weihnacht, Ray Weiss, Venicia Williams, Dr. Joseph Yang, ZETA Newspaper.

Both the book and documentary *Global Dumping Ground* would never have appeared were it not for those foundations that have supported the Center's environmental work during the four years of our hazardous-waste investigation. Our thanks go to the Mary Reynolds Babcock Foundation, Beldon Fund, C.S. Fund, The Educational Foundation of America, W. Alton Jones Foundation, HKH Foundation, Ruth Mott Fund, The Needmor Fund, The New World Foundation, Norman Foundation, Jessie Smith Noyes Foundation, Public Welfare Foundation, Ann R. Roberts, and the North Shore Unitarian Universalist Veatch Program.

Our deep appreciation goes to those at the various government agencies who went out of their way to help us. Of special note are the U.S. Marshals Service, Customs Service, Drug Enforcement Agency, Environmental Protection Agency, Federal Bureau of Investigation, California Highway Patrol, Los Angeles County Environmental Crimes Strike Force, and, in Taiwan, the Republic of China's Environmental Protection Administration. In these agencies and elsewhere, our thanks go to a number of people who provided invaluable assistance and must remain anonymous.

And, of course, our gratitude to the families and friends of the Center's staff, for their extraordinary patience and support throughout this long project.

Finally, a special thanks to an unnamed official in the North Carolina Department of Agriculture, who started it all. Years ago he called in a tip to the Center about a man named Charles Colbert, who was offering to buy up the state's stock of banned pesticides for a dollar a barrel for shipment to Latin America . . .

Credits

Global Dumping Ground: **The Book**

Editors:	Diana Hembree and David Kaplan
Consulting Editor:	David Weir
Preface:	Bill Moyers
Writers:	Sarah Henry, Constance Matthiessen, Dan Noyes, Eve Pell, Sharon Tiller, David Weir
Assistant Editor:	James Curtiss
Research Editors:	Stephen Levine and Loren Stein
Resource Guide:	Laura Lent
Graphics Coordinator:	Sue Ellen McCann
Index:	Barbara Newcombe
Research:	Juan A. Avila Hernandez, William Kistner, Miriam Ching Louie, Manfred Redelfs, Suzanne Rostler
Additional Research:	Glenda Anderson, Bill Gentes, Leslie Haggin, Melanie Haiken, Stella Ngai, Leora Romney, Elana Rosen, Robert Solley, Lisa Tsurudome
Editorial Assistance:	Theo Crawford

"Global Dumping Ground": The Documentary

Executive Editor and Correspondent:	Bill Moyers
Producer/Director for CIR:	Lowell Bergman
Executive Producer for KQED:	Ken Ellis
Executive Producer for CIR:	Sharon Tiller
Project Director:	Dan Noyes
Associate Producer:	William Kistner
Editor/Post Production Supervisor:	Herb Ferrette
Assistant Producer:	Juan A. Avila Hernandez
Production/Unit Manager for CIR:	Sue Ellen McCann
Unit Manager for KQED:	Nancy Carter
Camera:	Gino Bruno (Washington, D.C.); Walter Dumbrow (Colberts Interview); Carl Gilman (Los Angeles and Mexico); Norman Lloyd (Taiwan and China); Geoffrey O'Conner (New York and Atlantic City)
Additional Camera:	Tom Krueger (Newark Federal District Court); Brian Sewell (Brazil)

Sound:	Nick Blanchett, Benson Cheng, Steven Day, Martin Lopez, Robert Meleta, Caleb Mose, Richard Wargo
Lighting:	Mark Dumbrow, Mike Lapins
Production Assistants:	Miriam Ching Louie, Suzanne Rostler, Robert Solley
Researchers:	Melanie Best, Robert Buechler, Mike Kepp, Bob King, Laura Lent, Andrew Porterfield, Manfred Redelfs, Courtney Stewart
Consulting Producer:	Robert Richter
Animation:	Video Arts, San Francisco
Animator:	Nancy Casalina
Graphics:	Margaret McCall
Still Photographers:	William Kistner, Catharine Krueger, Jim Tynan
Executive Producer for FRONTLINE:	David Fanning

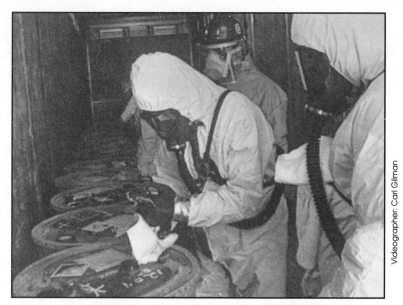

U.S. Customs agents checking barrels of waste.

ONE

The Paths of Least Resistance

To Sunday Nana, who owned a plot of land in the sleepy river port of Koko, Nigeria, it had seemed like a good trade. Each month, he received about $100—a small fortune in his village—in exchange for storing thousands of deteriorating metal drums on his land behind a rickety fence. As time went by, the drums popped in the heat, releasing acrid fumes, and many leaked, but Nana and his family continued to eat cassava that grew inside the fence. Despite the skull-and-crossbones labels on the drums, other villagers emptied some of them and carted them away to use as storage containers.

The drums, as it turned out, contained nearly 4,000 tons of toxic waste from Italy. Included in this waste were more than 150 tons of PCBs, or polychlorinated biphenyls, a compound linked to cancer and other diseases. Nigeria, which lacked an experienced waste management team, had no means of handling the materials. Once the situation became known, police simply sealed off the dumpsite.

Later on, there were reports that toxins from the drums had leaked into Koko's water, and that nineteen villagers had died after eating contaminated rice. Cleaning up the dump claimed other victims: three workers repackaging the waste suffered severe chemical burns, others vomited blood, and one man was partially paralyzed. Ultimately, the toxic

1

mess was sent back to Italy, where the mere unloading of the ship cost $11 million. The government of Nigeria, by now furious at what it called "toxic terrorism," threatened to execute anyone, native or foreign, convicted of importing hazardous substances in the future.[1]

Nigerian officials, who denied the reports of deaths from contaminated rice, said that Nana was told the barrels contained fertilizer. But Nana can no longer tell his side of the story; in 1990, two years after the drums were dumped on his land, he was dead. Nigerian officials claimed his death was due to respiratory failure unrelated to his exposure to toxic chemicals. Ironically, the former dumpsite on his land has been turned into a research center for the study of toxic waste.[2]

Were the Nigerian case an isolated incident, it would be bad enough. Unfortunately, it is just one of hundreds of cases of toxic-waste exports reported in recent years as industrialized countries have found themselves overrun with their own garbage. More than 2.2 million tons of toxic garbage cross borders each year, and there is no end in sight.[3] From used chemicals that illegal "midnight haulers" from the United States pour down ditches in Mexico, to lead battery wastes from Japan that poison smelter workers in Taiwan, what the developed countries throw away is making the world sick.

Some shipments threaten entire communities of unwitting victims: one U.S. company recently shipped ten tons of poisonous mercury waste to a British reprocessing plant in South Africa. Downstream from the plant, Zulu villagers now drink from a river with a mercury concentration 1.5 million times higher than the standard set by the World Health Organization. Said one American scientist, "If this were happening in the United States, all hell would break loose."[4]

A powerful economic equation drives the global trade in poisons: mounting piles of hazardous waste, a shrinking supply of disposal sites and exorbitant profits for people who can get rid of it—legally or illegally. Blocked by increasing government regulation and local opposition to disposal sites, the stream of waste constantly searches for new outlets. One ad in the *International Herald Tribune* put the matter explicitly:

> Thinking about making money? Hazardous toxic waste a billion-dollar-a-year business. No experience necessary. No equipment necessary. No educational requirements. Think of your financial future and call now for exciting details.

Both criminals and legitimate entrepreneurs sense handsome profits from this excess of hazardous waste, from steering a flow of harmful substances along the path of least resistance toward what they hope will be a final resting place. "I'd slash my wrists if I didn't think that there is enough greed in the world to find someone to take Philadelphia's trash," said one official of that city.[5]

All too often, however, the waste ends up in poor communities, migrating within the United States from the industrial Northeast to the more rural South; or in Great Britain, from England to Wales. Similarly, on the world stage, hazardous waste from the industrialized nations frequently has a one-way ticket to the developing world. Some Africans have even equated the traffic in toxic waste to the slave trade, although the direction has been reversed: the toxic substances that the industrialized world wishes to discard now flow to the developing world.

More than 3 million tons of wastes were shipped from the industrialized world to less-developed nations between

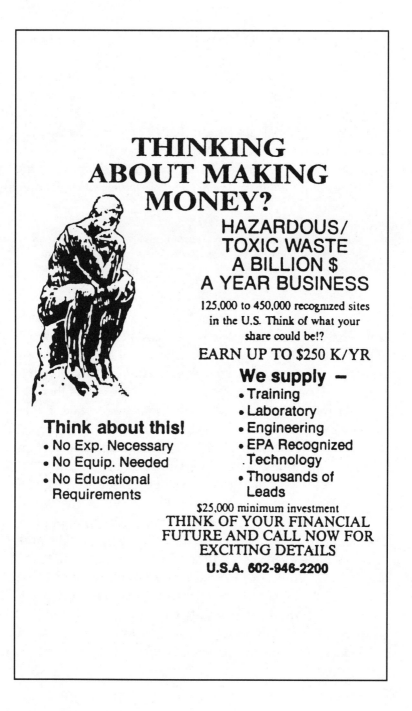

1986 and 1988, according to the environmental organization Greenpeace. Sometimes the deals were made with the approval of governments, sometimes not. The amounts of money to be earned from waste imports were so large that despite the health and environmental risks, some impoverished nations felt they could not refuse to enter this trade. The West African nation of Guinea-Bissau, for example, hoped to make $120 million a year—more than its total annual budget—by agreeing to store industrial wastes from other countries, until public protest over the hazards involved forced the government to back out. Asked why they had been willing to consider the deal, one government official said sadly, "We need the money."[6]

A series of odysseys in the late 1980s first drew worldwide attention to the issue of waste exports. Ships laden with hazardous wastes were refused admittance by country after country and, with their cargoes of poison still aboard, sent back to roam the seas.

The gravest danger to less developed countries, however, stems not from vagabond ships carrying deadly cargo, but from the legal, routine shipments of "recyclable" wastes: mercury residue, lead-acid batteries, and other refuse from which valuable materials are extracted by low-paid Third World laborers and then reprocessed or sold for reuse. This extraction often takes place in plants filled with choking fumes and lead dust, where workplace safety rules and enforcement are far less stringent than those in the First World. Both the workers and the people living near these factories are threatened as a consequence of this legal recycling trade.

Toxics: Made in USA

Although major waste-exporting nations range from Italy to Japan, the United States is by far the greatest producer, generating ten times as much hazardous waste as all of Western Europe. As a nation, America can't even keep track of all the hazardous waste it produces each year, officially estimated at 500 million tons.[7]

The United States, in fact, produces more than one ton of hazardous waste a year for each man, woman, and child in the nation. This ton is made up of the flotsam and jetsam of modern life, including the by-products of plastic, pesticide, and pharmaceutical manufacturing. No one outdoes Americans in the creation of both hazardous and non-hazardous wastes: While the average New Yorker generates four pounds of garbage a day, each Roman generates a pound and a half, and each resident of Manila a pound.[8] The aluminum the United States discards every three months could rebuild the nation's entire commercial air fleet.

Instead of being recycled at home, however, hundreds of thousands of tons of U.S. waste are destined for export, from computer batteries containing cadmium, which attacks the liver and kidneys, to industrial chemicals that can harm every organ in the body. While much of the world doesn't get to share in the United States' high standard of living, more and more of the world's people are sharing the burden of America's throw-away culture.

Meanwhile, as public consciousness of environmental hazards has increased, disposal within U.S. borders has become more expensive and difficult. Not only do landfills across the country leak, Americans have discovered, but incinerators spew out hundreds of tons of dangerous pollutants each year. Officials have taken note of how controversial the placement of toxic-waste facilities has become: a 1984 report commissioned by the California Waste Management Board recommended that incinerators not be located within five miles of a middle to upper-income community, noting that the residents least likely to offer political resistance would live in a low-income rural area.[9]

News coverage, however, has helped alert citizens to the potential hazards of toxic dumps. In the late 1970s, U.S. television showed terrified and angry families forced to evacuate their homes because of chemicals dumped into Love Canal near Niagara Falls in New York. Later reports of dioxin-contaminated waste oil on the streets of Times Beach, Missouri, underscored this danger. These and other well-publicized incidents gave rise to the NIMBY (Not In My Backyard) and the more generous NIABY (Not In Anyone's Backyard) movements.

In the United States, hazardous-waste landfills and incinerators are located disproportionately in poor and minority communities and, unlike some environmental concerns, the NIABY movement cuts across class and race lines. In New York City, a group of young Latinos called the Toxic Avengers fight what they call "environmental racism." Alan La Pointe, a community activist in Richmond, California, a low-income city near San Francisco, led a successful effort to stop construction of a massive waste-burning incinerator in his town. "The expenses were way out of line, and we were worried about poisonous substances getting into the air," he said.

As toxic residues pile up across the United States, the country has run out of places to put them. Hundreds of landfills have closed in the last decade; from 1984 to 1988, the number of operating landfills shrank from 1,500 to 325. At the same time, the costs of disposal soared.[10]

As recently as the late 1970s, U.S. manufacturers could simply bury wastes in landfills at little cost. In 1978, for example, auto makers could bury a ton of paint sludge for $2.50. Burning the stuff cost more: $50 a ton. By 1987, eleven years after the passage of the Resource Conservation and Recovery Act, the price of burial began at $200—if a landfill could be found in which to bury it—and burning the sludge could cost as much as $2,000 a ton.[11]

7

Faced with rising disposal costs, some manufacturers have taken a sharper interest in waste exports. Sometimes, such dumping may not even involve waste crossing a border. About 1,800 U.S.-owned manufacturing plants called *maquiladoras* are located just over the border in Mexico, where regulation of industry ranges from slight to none. Outside the plants, children play near open sewers that bubble with a chemical stew. Despite a 1983 agreement between the United States and Mexico requiring American companies in Mexico to return waste products to the United States, many of the factories have simply taken the most expedient course and poured chemical wastes down drains or into irrigation ditches, dumped them in the desert, or turned them over to unlicensed Mexican firms for disposal.[12]

"The *maquiladoras*, they use chemicals and they dump everything in the river," said Rosa Devora, who lives near Tijuana. "The wells are at the river's edge and that is where all the contamination ends up."[13] At night, she added, the *maquiladora* near her home discharges foul black smoke into the air and smelly chemicals into the river. Her two grandchildren, like others in the neighborhood, are continually sick with rashes, bronchial problems, and blisters.

"Worse Than Sending Guns": Toxics and Diplomacy

Besides posing a grave threat to human health and the global environment, the trade in hazardous waste has strained relations between the developed and developing nations. "I am concerned that if U.S. people think of us as their backyard, they can also think of us as their outhouse," said one official of an overseas environmental organization.[14] In 1972, a declaration adopted at the United Nations Conference on the Human Environment in Stockholm stated that each nation is responsible for ensuring that activities within its jurisdiction or control do not damage the environment of other states. Unfortunately, this stricture—called Principle 21—is not enforced.

Enforcement of environmental laws has never been simple. Former U.S. Environmental Protection Agency administrator William Ruckelshaus recalled: "In the late seventies when we began to write regulations for the control of hazardous-waste disposal . . . we didn't know where the generators were; we didn't know what was in their waste streams, or how much there was of it or how hazardous it might be; and we didn't know where it was going."[15] Although things have changed since then, enforcement remains a goal rather than a reality.

Indeed, U.S. environmental legislation, however well-intentioned, has actually encouraged the flow of waste out of the country. The 1976 Resource Conservation and Recovery Act marked the federal government's first involvement in regulating the disposal of hazardous waste. The act made the EPA set standards for landfills, decide the criteria used to label materials as hazardous, and mandate a "cradle to grave" system for tracking the waste. Most important, the "cradle to grave" provision forced the producer of such waste to assume financial liability for any damage it might cause in the future.

This legislation provided incentives for sending waste abroad, giving manufacturers a way to dodge their new open-ended liability. Company X, for example, might dispose of its used-up cadmium batteries in a U.S. landfill at considerable expense. But, several years from now, if that landfill should leak, the company would continue to bear responsibility for the cadmium and any health problems it might cause. However, if Company X pays a waste broker to ship it to China, the heavy metal becomes someone else's burden.

9

In the Carter administration, U.S. officials discovered and stopped one early attempt to ship poisonous waste abroad. After they learned of a Colorado company's plan to dispose of millions of tons of wastes in Africa, State Department officials warned that this practice could provoke anti-American outcries from other countries. African nations might "condemn the U.S. for dumping its wastes in the black man's backyard," said one official. In the early eighties, however, the transnational trade in waste grew swiftly. By 1983, a cargo of hazardous waste "crossed a national frontier more than once every five minutes, 24 hours a day, 365 days a year" within Western Europe and North America, according to the Organization for Economic Cooperation and Development. The trade spread, as industrialized nations like Canada, developing states like Taiwan, and less developed nations in Africa and Asia all earned hard currency by agreeing to import waste.

Some poorer European countries, including Romania and East Germany, also traded heavily in waste during the eighties. By 1989, as trade barriers fell across Europe, more Western countries began to talk about sending their hazardous wastes to Eastern Europe. Moreover, the need for hard currency may spur the development of hazardous industries and waste-disposal facilities in Eastern Europe: a polyvinyl chloride plant was recently proposed for the Ukraine and a petrochemical complex for Siberia.[16] Although Poland and East Germany quickly banned hazardous-waste imports, environmentalists worried that the new governments, desperate for hard currency, would allow waste to slip through.

What exactly is in the toxic stew that migrates across borders? Waste shipments include the detritus from the manufacture of such products as soap, plastics, pesticides, medicines, fertilizers, and explosives. Depending on the shipment, the waste may include substances like heavy metals, organic chemicals, dioxins, PCBs, and cyanide. In addition, many ordinary household products contain ingredients which, once discarded, become hazardous waste, including antifreeze (ethylene glycol), drain cleaners (hydrochloric acid), and floor polish (nitrobenzene). All of these may end up in municipal waste shipped overseas.

No one can say how much hazardous waste has been dumped on land or at sea or across borders, since this lucrative trade has yet to be fully inventoried, much less controlled. By all accounts, however, it is growing. While in 1980, only twelve notices of intent to export hazardous waste were issued by the U.S. EPA to foreign countries, as required by U.S. law, in the first half of 1988 the number jumped to 522.[17] The United States shipped 100,000 tons of hazardous waste abroad in 1987, according to official figures, but over the next two years, that figure rose by 40 percent.

Joel S. Hirschhorn, a former official at the U.S. Office of Technology Assessment (OTA), sees no letup. He blames our disposable culture, citing such "single-use products" as Styrofoam hamburger containers as well as innovations like throw-away video advertising cassettes designed to be used only two or three times. "The cassette will end up in an incinerator, generating air pollution," he said in disgust.

Hirschhorn added that the United States doesn't do a very good job of measuring how much waste is being generated within its own borders, "so it's a little naive to think we do a good job measuring how much goes to foreign countries."

Recent reports suggest the problem is out of control. A 1988 report by the U.S. EPA inspector general warned that tons of waste were being disposed of improperly—and that the agency did not even know how much waste was actually exported to other countries. EPA inspector general John C. Martin labeled the agency's hazardous waste export regulations "a shambles." Although a follow-up inspector general's report two years later found that the EPA had corrected many of those deficiencies, the hazardous-waste trade continues unabated.

Making matters worse, there is virtually no international machinery to monitor the waste trade, much less police it for violations. These conditions have helped attract illegal waste dealers, who work in a shadow industry comprising what one African magazine called "discreet brokers and intermediaries, complaisant shipping firms, and sundry ghost companies registered in places like the Isle of Man, Gibraltar, Liechtenstein, and of course, Switzerland."[18] One California company, for example, signed a multi-million-dollar deal with the Kingdom of Royaume de Tetiti in the South Pacific, which promised to accept shipments of hazardous waste, only to find that the Royaume wasn't a country but a group of uninhabited islands—and that the "king" was a notorious con artist wanted for questioning in Singapore and Thailand.[19]

Many industrial nations, furthermore, are unable to police even their legal waste dealers. The U.S. EPA, for example, cannot block a waste shipment if it has been accepted by another nation, even if the agency knows that the recipient country lacks the technology to handle it safely. Neither can EPA's Canadian equivalent, Environment Canada, no matter how harmful it might be. "Governments could fall because of the [toxic waste trade]," says Noel Brown of the United Nations Environment Programme. Exporting waste, he added, "is worse than sending guns, because it affects everyone."

The Basel Convention: Legalizing 'Toxic Terrorism'?

Although environmentalists and Third World leaders have urged a complete ban on exports of hazardous waste, the response has not been encouraging. Hopes rose briefly in

March 1989, when 117 nations sent representatives to Basel, Switzerland, to hammer out a treaty on the export of toxic waste. There, in a convention center decorated with abstract paintings that one observer described as "looking curiously like leaking barrels of waste," the delegates, some in business suits and others in tribal dress, argued heatedly over what should be done about the global waste trade. Outside, Swiss students and environmentalists dressed in the white space suits of a toxic waste clean-up crew rolled metal drums down the cobblestone streets to protest waste exports.

While all the African states and most Third World nations wanted an absolute ban on waste exports, the industrialized nations refused. The United States blocked any proposals for comprehensive regulation, such as requirements that exporting nations punish illegal waste exporters and take back wastes that could not be properly disposed of in the recipient country.

The main achievement of the Basel Convention, as the treaty is called, is that waste exporters must now notify and receive permission from waste importers before any shipment may proceed. The treaty also calls for signing nations to accurately label all international waste shipments, to prohibit waste shipments to nations that have banned them—and to try to reduce such exports to a minimum. It does not address waste exports intended for recycling.

While some signing nations hailed the treaty as a step forward, the African states and others protested that the agreement legalized a harmful practice. They worried that the less-developed nations have no mechanisms to enforce the provisions and therefore remain at the mercy of the First World. U.S. opposition doomed perhaps the most significant proposal: that no nation ship waste to a country with standards for waste disposal lower than its own. Environmental groups were equally distressed. Outside the Basel convention center, just before the final vote, Greenpeace-

Switzerland hung a bright yellow banner decorated with the words "Basel Convention Legalizes Toxic Terrorism."

Some industry groups, such as the U.S. Chemical Manufacturers Association, say they support the tenets of the Basel Convention, but lobby diligently to keep the waste trade alive. "Unnecessary restraints on exports, or requirements to handle them as hazardous wastes outside the United States, could seriously disrupt environmentally sound and important global commercial activities," according to the CMA.

Some progress toward limiting the trade has taken place, however. A year after the Basel meeting, an agreement between the European Economic Community and 68 African, Caribbean and Pacific (ACP) countries was reached, banning exports of toxic and radioactive wastes from those European nations to the ACP nations, with the exception of shipments to those states with "adequate technical capacity" for waste disposal. But many environmental groups still seek a complete ban on waste exports, echoing the view of a Common Market official who observed, "What we cannot dump safely in our own back yard, we should not allow to be dumped in somebody else's."[20]

Nigerian writer Chinua Achebe, whose novels chronicle the painful clash of traditional and modern cultures in Africa, agrees. "On our own continent, there are all kinds of mistreatment," he says. "The most recent is the dumping of toxic wastes from the industrialized world in Africa. . . . You have examples of officials who take money so that toxic waste can be dumped in their territory. It's impossible to contemplate that kind of situation without being very, very bitter." But Achebe is guardedly optimistic about the possibility for change: "These things would not go on if the people said no. . . . When the people become well educated and well organized, then they will be able to say no."[21]

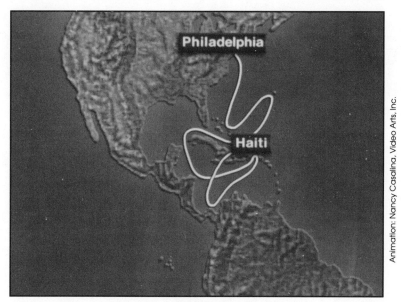

Odyssey of the *Khian Sea*.

TWO

Odyssey of the Poison Ships

W hen Joseph Paolino, a Philadelphia waste contractor, struck a deal to get rid of 200,000 tons of ash from the city's garbage incinerators in early 1986, he clearly believed there was money to be made. The municipal docks were piled high with huge mounds of the gray refuse, lining the Schuylkill River like a vast dunescape. But as more and more landfills around the city closed down and citizens began to protest the gritty ash blowing through their neighborhoods, Paolino realized he had a problem.

Frustrated and faced with a massive backlog of the toxic ash, he then came up with what seemed a brilliant solution—hire a ship and find a dumpsite overseas.

That ship, the *Khian Sea*, would create an international scandal as it hauled Philadelphia's incinerated trash around the world. The ship was to spend twenty-seven months wandering the seas like a gypsy, with "more lives than a cat and more names than a Spanish duke," as one reporter put it.[1] From Panama to Haiti, Guinea-Bissau to Sri Lanka, one government after another refused to accept the ship's load of poisonous garbage. Before this bizarre odyssey came to an end, the ship's name had changed from *Khian Sea*, to *Felicia*, and finally to *Pelicano*; it flew the flags of both Liberia and Honduras. Along with a maze of lawsuits and investigations, the ship also sparked action in the U.S. Congress and

17

edicts from half a dozen other nations. One congressman declared the story of the *Khian Sea* "better than a mystery novel."[2]

The *Khian Sea*, however, was but one ship in a seeming armada of toxic vessels unleashed by the developed world in the late 1980s. People from Philadelphia to Djibouti were outraged by the stories of these "poison ships"—freighters laden with hazardous waste, crisscrossing the Caribbean Sea and the Atlantic and Indian oceans looking for a home. The waste came not just from the United States, but from a string of industrial nations: Belgium, Canada, Italy, Switzerland, and West Germany. Like a modern *Flying Dutchman*, that legendary ghost ship condemned to sail forever without making port, the toxic ships came to represent better than any other symbol the growing global crisis in hazardous waste.

The ships came christened with exotic names that belied their unsavory cargos: *Zanoobia, Jolly Rosso, Karin B, Lynx, Pro Americana, White Stork.* Instead of wheat or oil, lumber or manufactured goods, their holds were filled with the refuse of modern life, from barrels of toxic chemicals to tons of dioxin-laced incinerator ash.

In one elaborate scheme, three different ships carried a single load of 2,200 tons of toxic material to four continents over a period of fourteen months. The courses they charted ranged from Italy to Djibouti, then on to Venezuela and Syria. Branded "barrels of death" by the European press, the cargo consisted of 20,000 containers of pesticides, PCBs, cyanide, paint solvents, and dioxins.

Faced with soaring costs for disposal in the West, both government agencies and private industry struck deals with waste haulers to ship industrial refuse to developing countries desperate for hard currency. Tempted by the quick profits to be made and often unaware of potential environmental and health hazards, countries like Panama, Haiti, Guinea-Bissau, Benin, Peru, and Argentina initially approved lucrative waste-dumping deals. Some of the officials in these countries believed shippers who told them that the toxic cargos were in fact harmless.

But Greenpeace activists and others launched a worldwide campaign to warn countries about this toxic trade, declaring it "morally reprehensible to force developing countries into this choice between poison and poverty."[3] As foreign newspapers picked up the story, one country after another began to turn the ships away. In Nigeria, when government officials discovered that Italy had dumped toxic waste at one of the country's river ports, the reaction was swift and harsh: future perpetrators would face the death penalty.

Over a period of more than two years, these rusting, reeking freighters circled the globe like buzzards. The world press branded them "ghost ships," "Flying Dutchmen," "ships of fools," "leper ships," "gypsy vessels," and "vaisseaux du poison." International outcasts, the ships changed crews and captains, owners and flags, in a futile attempt to outwit increasingly irate citizens and public officials around the world. No ship, however, caused more outrage than the *Khian Sea*.

The Saga of the *Khian Sea*

In America, the strange story of the *Khian Sea* began with a problem all U.S. cities face: what to do with the garbage generated by a disposable society. Philadelphia had decided to burn its trash, turning tons of municipal garbage into mountains of toxic ash. From the mid-1970s to 1988, municipal incinerators burned 40 percent of the city's garbage, generating some 2,000 tons of ash each day.

As of 1985, the city was relying on nearby New Jersey landfills to dispose of the waste. As all of the landfills in and around Philadelphia began to fill up and close, however, the city contracted with waste handler Joseph Paolino and Sons in November 1985 to dispose of up to 200,000 tons of the incinerator ash at a cost of $41.74 per ton. After a series of unsuccessful attempts to unload some of the ash in Virginia and South Carolina, Paolino finally succeeded in dumping part of it in the Dispose-All Refuse landfill in Columbia County, Ohio.

Ohio citizens living near the landfill protested loudly, however, claiming that the foul-smelling ash was making them sick. Despite assurances from the local EPA that the ash did not pose a public health risk, state tests revealed that it was laced with lead, chromium, arsenic, and traces of a deadly dioxin compound.

An EPA report released much later indicated that the material was in fact quite toxic.[4] According to the report, which was authored by EPA Inspector General John C. Martin, the ash contained levels of dioxins (up to 4.7 parts per billion) higher than those found in Times Beach, Missouri—the town that had to be evacuated in 1983 —and high levels of such toxic metals as lead and cadmium. Thwarted in their attempts to dispose of the ash domestically and faced with a monumental backlog, Paolino and Sons finally began to explore the possibility of shipping the waste overseas. Philadelphia had apparently discussed its first export proposal with a representative of the government of the Bahamas as early as the spring of 1986, according to 1988 Congressional testimony.[5] Although newspaper accounts would later tell a different story, a city official maintained that the Bahamian official raised no objections to disposal in his country.

It was this ill-fated business deal that launched the twenty-seven-month voyage of the *Khian Sea*. In June 1986, Paolino and Sons subcontracted with Amalgamated Shipping Corp. to transport more than 13,000 tons of ash to the Bahamas for $26.75 per ton, allegedly with assurances from

Amalgamated that they had official permission to unload the cargo. Two months later, the Norwegian-owned ship flying the Liberian flag sailed for Ocean Cay in the Bahamas, where that country's Ministry of Health emphatically denied it permission to unload.

Over the next year, with virtually no publicity or attention from government authorities, the *Khian Sea* cut a swath from Florida, past Puerto Rico and the Dutch Antilles, to Central America. The ship wandered the high seas while Amalgamated looked unsuccessfully for a place to unload Philadelphia's garbage. Meanwhile the controversy over the ship gathered momentum. In a Gordian knot of charges and counter charges, the city of Philadelphia claimed it had no control over the *Khian Sea*.

Nonetheless, it withheld payment of more than half a million dollars to Paolino and Sons for failing to provide proof that the ash had been disposed of legally. Paolino then filed suit against Amalgamated in March 1987 for fraud and breach of contract: not only had the company failed to get permission to unload the ash in the Bahamas, it had sought to negotiate deals in the Dominican Republic, Honduras, Guinea-Bissau, and possibly other countries, and failed to get dumping approval from any of them.

Bruce L. Phillips, Paolino's attorney, claimed the company made repeated requests to Amalgamated that the vessel return to Philadelphia so the ash could be unloaded and trucked to a domestic landfill site. "Defendants have advised plaintiff that the vessel *Khian Sea* is 'off the coast of Florida' or 'somewhere in the Caribbean Sea,' but have refused to disclose its exact location despite demand," the lawsuit alleged.[6] Amalgamated then responded by counter-suing Paolino for the $403,000 owed them for carting off the ash in the first place.

Until now, the *Khian Sea* odyssey had attracted only local coverage, but a stepped-up campaign by Greenpeace, combined with national U.S. television reports, eventually sparked worldwide interest in the ship. In October 1987, when the *Khian Sea* attempted to dock in Panama, it became the first toxic vessel to receive international attention. What led to the explosion in interest was actually an unrelated plan by a second Philadelphia waste contractor, Bulkhandling, Inc.

While the *Khian Sea* was the public target of fierce attacks by Panamanian officials, it was in fact a U.S. EPA report about this new proposal from Philadelphia that alerted Central Americans to an impending environmental time-bomb. Bulkhandling planned to dump 250,000 tons of incinerator ash a year on Panama for a roadbuilding project that presented "a significant potential danger to human health and the environment," according to EPA's inspector general John C. Martin. Not only was the area of the road project dotted with pristine beaches and sparkling blue lagoons, the road itself was to be built through a fragile wetlands area that was Panama's main breeding ground for the white shrimp, the country's main food export. On a December 1987 "CBS Evening News" broadcast, the port captain in Bocas del Toro, Panama, declared that the *Khian Sea* would be "arrested" if it attempted to enter the harbor. Rejected once again, the ship set off in search of another possible port of entry.

By November 1987, the *Khian Sea* had logged more than seventeen months at sea and had been rejected by the Bahamas, Bermuda, Honduras, Panama, and the West African nation of Guinea-Bissau, when the ship finally arrived in Haiti. Under cover of recent election turmoil, Amalgamated had struck a deal with a company called "Cultivators of the West," operated by Antonio and Felix Paul, brothers of Colonel Jean-Claude Paul (former leader of the notorious paramilitary organization, the Tontons Macoutes). The brothers had secured an import permit for the *Khian Sea* from the Haitian Commerce Department to unload the

Video: WCAU-TV, Philadelphia

**The *Khian Sea* docked in Haiti,
New Year's Eve 1987.**

ship's cargo of "fertilizer" at the port of Gonaïves. A few months later, in March 1988, Jean-Claude and Antonio Paul were indicted in the United States for links to another suspicious business operation, cocaine smuggling.

The ship sailed into Gonaïves harbor on New Year's Eve, where the ash was to be barged ashore and used as fill, according to Roger Duggan, commercial attaché at the U.S. Embassy. When he visited the site where the *Khian Sea* was to unload, Duggan discovered a vessel eaten away by its corrosive cargo. "The ship looks like a derelict," he said. "The exterior is completely rusted. It may have been painted once, but the paint's gone."[7] By this time, the pariah ship was attracting the attention of the local media and a Greenpeace camera crew, so Robert Dowd, traveling on the vessel as a representative of Amalgamated Shipping, was determined to show there was no cause for concern. Pinching a sample of the ash from a mound lying on the dock, he put it into his mouth and chewed. Peering out from under a baseball cap with New York emblazoned on it, Dowd said, "See, that's how worried I am of its toxicity."[8] Later Dowd's father Henry, the vice-president of Amalgamated, added: "He's my son, all right—crazy ex-marine—he ate the ash."[9]

The younger Dowd's appetite for incinerator ash apparently failed to impress the Haitians, because the newly elected civilian government announced a ban on foreign waste ships entering Haitian waters and ordered the *Khian Sea* to leave. Speculating on what had soured the deal, Henry Dowd complained later: "People in Haiti were happy about it—we paid double the local salary to unload it. They thought they would have long-term employment, but false press destroyed the whole thing. Instead of being a good deal, it turned into a nightmare."

The Caribbean Conservation Association, a regional environmental group whose members include the governments of seventeen island nations, reacted angrily to the

Video: WCAU-TV, Philadelphia

Robert Dowd samples the ash from the *Khian Sea*.

episode. "We are nobody's outhouse," said the association's executive director Michael King. "If they have so many acres of land in the United States, why should they dump their garbage on other people?"[10]

So adrift again, the *Khian Sea* sailed out of the Gonaïves harbor in February 1988 with 10,000 tons of Philadelphia's blacklisted incinerator ash still eating away at its cargo bays, but only after it had managed to unload some 3,000 tons before being ordered to leave. Responding to angry demands from Haitian officials, the EPA sent a team from Washington to test the ash and advise on its containment or removal. Although the EPA study concluded the ash did not pose a threat to public health, it recommended workers use protective equipment when working near the "contaminants."

At the same time that the *Khian Sea* was departing Haiti, another ship, the *Bark*—registered in Norway and under contract with Paolino's competitor Bulkhandling—was leaving Philadelphia with another 15,000 tons of the city's ash. The initial plan was for the ash to be used in constructing a roadbed in Panama. But, armed with Greenpeace and EPA studies showing the ash contained lead, mercury, aluminum and organic products that form dioxin, Panama's Ministry of Health denied permission for the ash to enter the country. "Panama will not accept it for the same reason six states in the U.S. will not accept it," said a spokesman for the ministry. "If it is not good for the United States, neither can it be any good for Panama."[11]

Next the *Bark* turned up in the West African nation of Guinea, where the ship's reception at first seemed favorable. In the first reported foreign disposal by any one of the "poison ships," Bulkhandling announced on February 4, 1988, that it had disposed of the ash for $40 per ton on the Guinean island of Kassa near the capital city, Conakry. The waste material was now to be used, not as fertilizer, but for constructing concrete-like bricks. But the deal sparked an international furor when it was revealed that the *Bark* had violated a two-year prohibition against all foreign wastes entering that country. The *Bark's* Norwegian owner only agreed to remove the contraband waste after a diplomatic brouhaha, in which Guinean officials arrested the Norwegian consul-general for complicity in the dumping. After Bulk-handling arranged for the return of the ash to the United States, the diplomat was released.

Meanwhile, by late spring of 1988, the odyssey of the *Khian Sea* had entered its twentieth month, when, rusted and radarless, the ship limped back into Delaware Bay. The letter "K" was still visible on its faded blue smokestack, as it returned to an anchorage just 100 miles from the city where it had first set off nearly two years before. By now, the 400-foot-long vessel had been spurned by at least seven countries, and Amalgamated apparently hoped to return the blighted load of ash to the city that had generated it.

Once again fate was ill-disposed toward the *Khian Sea*: a suspicious two-alarm fire damaged the ash-hauler's pier in Philadelphia, making it impossible for the ship to dock and unload. A veteran river pilot guided the battered *Khian Sea* to an anchorage some eighty-five miles and six hours from Philadelphia, where it was ordered to wait for a Coast Guard inspection and assessment by the local EPA before moving up river. The inspection crew that boarded the ship later that morning to check its navigation, propulsion, firefighting, and steering systems called it "a rust bucket for sure, a garbage pail."[12] They also said it smelled of burned trash about its decks.

If that were not trouble enough, legal squabbles between the original contractor and the shipping firm led Paolino and Sons to refuse to allow the *Khian Sea* to dock unless Amalgamated pledged to clean up the ash it had dumped back in

27

Gonaïves, Haiti. According to Paolino's lawyer, the shipping firm would not be paid by the city unless they could show that the ash had been disposed of legally.

International Fugitive

Caught in this catch-22 between the city and Paolino, under cover of darkness on May 28, 1988, the *Khian Sea* once again steamed out of Delaware Bay toward the open sea. The ship had dodged U.S. Customs, disregarded state safety regulations, and ignored orders from the Coast Guard to remain anchored because neither its sonar nor its radar was working. Edward Roe, the Coast Guard port captain in Philadelphia at the time, described the ship's condition as questionable. "They were not sure it could move on full power, and there were a series of instruments out of order."[13] When asked why the Coast Guard did nothing to intercept and turn the ship back, Roe said it would virtually take a cannon shot across the bow to stop a ship of that size. The Coast Guard would only take belligerent measures in extreme circumstances, and shady trafficking in incinerator ash was apparently not one of them.

After escaping the Coast Guard, the renegade ship headed for Africa, where it next attempted to dump its load in Senegal, the Cape Verde Islands, and Guinea-Bissau. The Organization for African Unity—already incensed over the hazardous-waste disaster in Koko, Nigeria—passed a resolution condemning the use of the African continent as a dumping ground. Nigerian President Ibrahim Babangida declared, "No government, no matter the financial inducement, has the right to mortgage the destiny of future generations of African children."[14]

Driven away by Panamanians, Haitians, and now Africans, a fugitive from U.S. lawyers and the Coast Guard, the

Khian Sea sailed for Bijela on Yugoslavia's Adriatic coast. There it underwent repair and was renamed the *Felicia*, but Liberia cancelled registry on the vessel based on its refusal to obey Coast Guard orders to remain in Delaware Bay. Amalgamated then sold the newly christened *Felicia* to Romo Shipping Inc., which obtained permission to fly the Honduran flag. Greenpeace continued to track the chameleon ship through the Suez Canal and reported its attempts to dock in Sri Lanka, Indonesia, and the Philippines. In every country, however, officials made clear the refugee ship was unwelcome.

By now the ship had been renamed again, as the *Pelicano*, by its new owner Romo, which was clearly determined to end the ship's long voyage. In November 1988, the ship suddenly appeared in Singapore without its cargo. And no one—not its owner, its captain, or its crew—would say what had happened.

Greenpeace maintained that the ship had dumped the ash somewhere in the Indian Ocean, but Captain Arturo Fuentes told reporters visiting the ship that the *Pelicano* unloaded the unwanted cargo in a port, although he refused to say which one. Romo also declined to say where the ash went. Despite an investigation by the U.S. Justice Department, the mystery remained unsolved.

Although former *Khian Sea* official Henry Dowd of Amalgamated also claimed he did not know what had happened, he did speculate: "It was possible to dispose of it in India or Pakistan. They didn't want any publicity—just do it quietly in a small port." He said that he doubted the ash would have been dumped in the ocean, because it was dangerous to the crew and ship to open the hatch at sea. It would also mean losing insurance protection. "That's the reason we never considered it," he said.

Perhaps most disturbing is how little waste the *Khian Sea* actually represented. The Homeric odyssey that had lasted more than twenty-seven months, touched four continents, and set off a series of international furors had in the end involved no more than one month's worth of incinerated waste from one U.S. city.

Jelly Wax and the Ship *Zanoobia*

Although the United States produces more hazardous waste than any other industrial nation, it has not been the only country to engage in the bizarre high-seas dumping trade. One of the most elaborate disposal schemes of the 1980s originated in Italy, where a Milan-based firm, bearing the odd name of Jelly Wax, shipped 20,000 barrels of highly toxic waste from Northern Italy to the tiny African nation of Djibouti.

The barrels, containing everything from pesticides and paint residues to PCBs and dioxins, were collected from industrial sites throughout Italy and from the U.S. military. According to court documents filed later in civil and criminal cases, the waste generators included, among others, the U.S. air base in Aviano, Italy, the multinational aerospace corporation Morton Thiokol, and industries controlled by the Italian government. Francesco Rizzuto, an Italian lawyer representing the Syrian captain of the *Zanoobia*, Ahmed Tabalo, offers a wildly different scenario. He maintains that this scandal extends to links with the Medellin Cartel and the wife of former Romanian dictator Nicolae Ceaucescu.[15]

In February 1987, some five months after the *Khian Sea* steamed out of Philadelphia's harbor, the *Lynx,* a ship sailing under the Maltese flag, left the Italian port of Marina di Carrara, loaded with what came to be called "barrels of death," bound for Djibouti on the Horn of Africa. Once there, the wastes were to be buried in the desert forty kilometers from the city, according to a contract worked out between Jelly Wax and Djibouti's government. The deal had been brokered by a middle man, Gianfranco Ambrosini, the director of the Swiss-based waste management firm Inter-contact S.A.

Like the *Khian Sea*, the *Lynx* was fated not to make this first port of call, because local authorities believed, on account of erroneous reports from Greenpeace, that the cargo was radioactive. From Djibouti, the ship sailed for Venezuela, where Jelly Wax had made a deal through a Panamanian firm, Mercantil Lemport S.A., to unload the barrels in Puerto Cabello, a sleepy little town 100 kilometers west of Caracas. Puerto Cabello officials had a reputation for accepting anything—without official papers—as long as the bribe was large enough.

After the barrels had sat in the intense summer heat in Puerto Cabello for six months, however, they began to leak and allegedly cause health problems among nearby residents. When the press reported that two children had become very sick after playing near the dump, the Venezuelan government ordered Jelly Wax, in 1987, to remove the waste.

Two months later, a Cypriot-flag ship, the *Makiri*, arrived to pick up the rotting barrels, with orders to return the waste to Italy. But in the meantime, Jelly Wax had brokered yet another deal, this time selling the cargo for $200,000 to a Syrian businessman, Mohammed Samin. After port authorities in Cagliari, Sardinia, refused to grant the *Makiri* permission to dock, the vessel headed for Tartous, Syria, where the leaking barrels were stashed somewhere in that Syrian port in December 1987.

When they learned of this attempt to unload the lethal cargo in their country, angry Syrian officials ordered the immediate removal of the waste. So in the final transfer of these toxic chemicals, the Syrian-flag ship *Zanoobia* retrieved the drums of waste and headed back to Italy, where the saga had begun fourteen months before.

31

As of mid-1990, negotiations continued over who would be responsible for disposing of the waste—the Italian government, the owners of the *Zanoobia*, or Castalia, the company put in charge of unloading the poisonous barrels two years before.[16] Meanwhile, the ship remained at anchor near Genoa, pending the outcome of criminal and civil trials naming as defendants Jelly Wax, Ambrosini, and the Italian ministries of the Environment, Merchant Navy, and Foreign Affairs, among others. The crew, for their part, reportedly suffered from headaches, nausea, and severe skin irritation. Italian dock workers, after refusing for several months to unload the hazardous cargo, finally relented in late 1988. The barrels were repackaged and stored on a floating dock within a mile of the *Zanoobia*.

The international uproar and notoriety of the "poison ships" appears for now to have put the brakes on this deadly trade by sea. There are new restraints on trade imposed by individual countries and the United Nations-sponsored Basel Convention. Nonetheless, comments by the contractors responsible for the *Khian Sea* and the *Zanoobia* raise serious doubts about whether legislative initiatives will indeed halt this practice.

Maurizio Ambrosini, head of the waste firm working with Jelly Wax, expects few changes. "I don't have much faith in Italian politicians," he said. "All you have to do is wait for the storm to pass and then everything will go back to the way it was before."[17]

Henry Dowd, the contractor responsible for the *Khian Sea*, is also skeptical: "There probably are other *Khian Sea*s around the world taking municipal ash and waste," he suggested. "They are a lot smarter than we were and can avoid publicity. Hazardous waste . . . is similar to the drug trade. People can make a lot of money."

**AID emblem on barrel
in Zimbabwe.**

THREE

The Pioneering Colbert Brothers

In April 1983, a mysterious fire swept through an aging warehouse in a residential neighborhood of Newark, New Jersey. Firefighters dowsing the blaze soon realized that this was no ordinary fire: the four-story warehouse, they discovered, was bulging with several thousand drums of illegally stored toxic chemicals, including an explosive mixture of acids, solvents, gases, and pesticides.

To investigators from the state Department of Environmental Protection, the case at first seemed routine. New Jersey had its share of toxic-waste wheeler-dealers, and this appeared to be yet another abandoned waste site. But the story behind the warehouse turned out to be anything but ordinary. The trail of the two owners—brothers Jack and Charlie Colbert—led to a web of toxic-laden warehouses and businesses stretching across at least eight U.S. states and, ultimately, to shipments of waste to more than 100 countries, most of them in the Third World.

The Colbert brothers, it turned out, were pioneers. Much of their chemical export business was legal, but some of it, according to prosecutors, was not. By stuffing some of America's deadliest poisons into misleadingly labeled barrels and crates, the two ambitious New York businessmen became the first known entrepreneurs in the fraudulent export of hazardous waste.

35

The Colbert operation was a classic con game, say U.S. officials, a multi-million-dollar scam with a toxic twist. The brothers discovered that they could take barrels of chemicals banned by the U.S. government for sale in this country or classified as hazardous waste and turn them around for sale overseas at a smart profit. To the Third World they offered everything from banned pesticides such as DDT and chlordane to barrels of flammable solvents and acids. Their inventory even included rolls of lead-contaminated paper, obtained from the U.S. Treasury Department's Bureau of Engraving, that one unsuspecting buyer wanted to sell to an African country as toilet tissue.[1]

Back home in America, few of the corporations and government agencies that supplied the Colberts seemed interested in where their hazardous waste was going. Court documents reveal that the bulk of the brothers' business came from the U.S. government itself: the Colberts would contract with the U.S. Navy, for example, to ship its toxic wastes, surplus chemicals, and banned pesticides to one of the brothers' warehouses, often for as little as $1.00 a truckload. The Colberts would then try to find Third World buyers for the waste, which was often falsely labeled as pure chemical product. "The Third World was their hook," said New Jersey Department of Environmental Protection investigator Gary Allen. "They relied on no one complaining."

For such large-scale waste trafficking, the brothers needed help. Allen and other investigators looking into the Colberts' operation identified at least three waste-hauling companies as collaborators. These companies, licensed by the EPA, were accused of illegally transporting poisonous waste to the Colberts' Atlantic City warehouses, where it awaited export to Africa, Asia, and Latin America. According to New Jersey investigator Bruce Comfort, who worked with Allen, "Everybody was milking the hell out of the system."

Former Assistant U. S. Attorney James DeVita, who was in charge of prosecuting the Colbert case, described the brothers as the most notorious toxic-waste "dumpers" in U.S. history. "The Colbert brothers," he said, "were to toxic waste what the James brothers were to bank robbery."

The Colberts Speak Out

A conversation with the Colberts offered a rare glimpse into the dimensions of the international waste trade—and loopholes in U.S. environmental law. After turning down requests for interviews for more than three years, in March 1990, Jack and Charlie Colbert finally agreed to talk on camera for the taping of "Global Dumping Ground." By then, both were serving time in federal prison for fraudulent business practices in shipping waste overseas.

Despite their convictions, the Colberts remained adamant about their innocence. "Basically, we were involved in an innovative business," claimed Charlie Colbert, a divorced 43-year-old State University of New York law school graduate who grew up with his brother in a middle-class neighborhood of Riverdale, New York. "We were, in a sense, innovators ahead of the times because what you had was a whole definition in the environmental area that isn't really defined yet. . . . So we were in a gray area, and we were in an area that the society needs."

Jack and Charlie, dressed in khaki prison uniforms, sat in a locked visitors' room at the Otisville, New York, federal prison to be interviewed. In a rapid-fire New York accent, Charlie defended their toxic-waste exports: "You need experimentation in order to find out intelligent solutions for problems. . . . And we were the victims of that. We're basically pioneers in . . . the surplus chemical business, which is something that's a necessary business for the society."

37

Videographer: Walter Dumbrow

Jack and Charlie Colbert.

The brothers insisted their business was legitimate, and much of what they did was in fact legal. Federal regulations were loose enough so that almost anyone with a trailer truck and enough cash could purchase the government's un-wanted chemicals.[2] It is also perfectly legal to sell restricted poisons like DDT overseas. But relabeling the hazardous waste as "surplus chemicals"—as the Colberts did—was illegal.

Legal or not, the entrepreneurial brothers made a lot of money from these activities. "We did between $8 million and $10 million a year for fifteen years," Charlie said, check-ing off the amounts on his fingers as he recalled his lost fortune. "And some years we did as much as $15 million or $20 million. So you're looking at, conservatively, $80 mil-lion, and, on the high side, maybe as much as $180 million."

The brothers' first deal came about almost by accident. Charlie teamed up with Jack, the younger of the two, to begin an import/export business in the early 1970s. "What happened," said Charlie, "was . . . somebody called me up and gave me five truckloads of material from a military base, and we ended up selling it for $80,000. . . . That's why we were in the surplus chemical business."

As far as the Colberts were concerned, they were provid-ing a useful and lucrative service. "It was symbiotic for both us and the society," Charlie argued. "It was beneficial to us because we found a way to be competitive, and it was bene-ficial for the society because we were helping solve a prob-lem. Instead of chemicals going in the ground and costing a lot of money for disposal, they were being reused a second time."

Jack, who for most of the interview sat slumped over in his seat, straightened up suddenly to interrupt his brother. "They were not being reused a second time," he declared. "They were being used. They had never been used the first time. That was the whole thing. I mean if material isn't used, okay. It's virgin material still in the drum. Why bury a drum of good product?"

To find these "drums of good product," the Colberts placed an advertisement in trade publications, requesting chemicals that "are either excess inventory for you . . . or are no longer approved by the EPA in this country." The ad urged agencies to "call us before you dispose. . . . We can solve your recycling problems. . . . Open up your connections to the whole world for buying and/or selling."

It would seem this ad might have set off alarm bells in the EPA, which is, after all, mandated by law to track hazardous waste from "cradle to grave."[3] But this responsibility ends, unfortunately, at U.S. borders: the EPA has no legal authority to stop waste headed for an unsafe "grave" if that grave is in another country.[4] And by labeling their waste as a chemical product, Jack and Charlie Colbert were able to ship it overseas for nearly seven years before the federal government started asking questions.

During that time, the supply available to the two brothers of old, banned, and unwanted chemicals seemed limitless. After all, who would want to pay to bury pesticides like DDT in an expensive waste-disposal site when the Colberts were willing to buy them and haul them away?

One of the suppliers to the Colberts was the North Carolina Department of Agriculture, which sent the brothers DDT, chlordane, and other banned or restricted pesticides.[5] Federal agencies, including the U.S. Army and Navy, were also steady customers.

According to New Jersey state investigator Allen, at least 50 percent of the chemical waste found in one run-down, abandoned warehouse "was federal stuff." Court documents later traced most of these chemicals to the Navy Submarine Base in Groton, Connecticut, and the Norfolk Naval Shipyard in Virginia.[6]

And there were other well-known suppliers whose surplus ended up as part of the Colberts' transactions. "Just remember one thing also," Jack Colbert declared. "Most of the people we're buying material from, they weren't fly-by-nights: Ford Motor, Exxon. Aside from the government, most of the major corporations, Dupont . . ."

"Celanese," added Charlie.

"They all sold us material," said Jack.

From Jack Colbert's perspective, he was a middleman providing a service to corporate America. If a company produced 10,000 gallons of paint that was a little off specification, he explained, he and Charlie could sell the unwanted product overseas. "What I've done is marketing," he reasoned. "See, Dupont and a lot of these companies could have sold the material overseas themselves. But the thing is they're not set up with the marketing for it. It was legal to sell the material. But they don't want to invest the marketing dollars."

Marketing paint in the Third World is one thing, marketing toxic chemicals and waste is another. Here, the Colberts highlight one of the key issues of global dumping: under U.S. law, they could freely sell chemicals overseas—to any interested buyer—that the EPA considered too dangerous for use in the United States. Although many Third World countries are poorly equipped to handle and dispose of hazardous chemicals, officials in Washington have failed to restrict the "dumping" of banned chemicals in other countries' backyards.[7] It was not surprising, therefore, that the Colberts felt they had the tacit approval of the U.S. government to ship poisons overseas.

"The EPA in the United States banned DDT for use in the U.S., they didn't ban it for production in the United States," argued Jack Colbert. "There are a lot of products that were not approved by the EPA that are still in use in the entire world. Now . . . you're gonna say that our EPA knows better than 165 other countries in the rest of the world?" he asked incredulously. "Is that what you're saying?"

U.S. laws that permit the United States to ship its banned chemicals to poorer countries lead to a kind of "us vs. them" attitude reflected in the Colberts' arguments. "What do you want to do?" Jack asked. "Do you want it to be buried in America or do you want it to be sold in a Third World country? . . . Which would you prefer?"

Questioned about issues of personal responsibility, Jack asked rhetorically, "Am I sorry that I sold chemicals to the Third World? . . . No."

The Zimbabwe Deal

In the end, it was not the sale of banned chemicals but a deal involving U.S. taxpayers' money that proved to be the Colberts' downfall. In response to an advertisement in a major U.S. trade catalog, the Colberts received an order in 1984 from a company in the southern African nation of Zimbabwe. The order, from a firm called Chemplex Marketing, was for $54,000 worth of dry-cleaning fluid and degreasing solvents.

But what Chemplex received from the Colberts would hardly clean suits. The company in Zimbabwe was shipped 228 fifty-five gallon drums filled with what U.S. officials would later call "an unusable, recycled mixture of chemicals which contained highly toxic waste material."[8] The drums came from one of the Colberts' regular suppliers, Alchem-Tron, Inc., a waste-disposal operation in Cleveland, Ohio, with its own history of fines and fires.

Protesting his innocence in the deal, Jack Colbert put the blame on the Cleveland company. "I paid Alchem-Tron different prices for different drums and no matter what I paid Alchem-Tron, I got the same thing. I got garbage. I mean they took everything they didn't know what to do with and they sold it."

After paying sixty cents per gallon for this "garbage," the Colberts turned around and sold it for $2.60 per gallon to Chemplex in Zimbabwe. What the Colberts did not know was that the company had obtained the funds to pay for the garbage from a U.S. foreign-aid program—American taxpayers' money, in other words. This turned out to be the key to their prosecution in a federal criminal case.

What happened was this: The program in question was operated by the Agency for International Development (AID), and its officials were alerted to the deal when they received an angry overseas call from Chemplex. When the shipment from the Colberts arrived, the company complained it was not the pure solvents specified in the contract, but barrels filled with unusable poisons—some of them leaking.[9]

This was not the first time that a foreign company had complained to the United States about one of the Colberts' deals, but it was the first time the U.S. government had paid the bill. Sitting in an office in New York, Al Rossi, a lanky AID investigator assigned to the case after Chemplex complained, pointed to a photo of the rusted and crumbling barrels he found in Zimbabwe. "As you can see in the photograph, the barrel appears to have been eaten out from the inside," he said. "That's to say the material, in addition to being toxic, was very corrosive."

The barrels, filled with a potentially explosive mixture of recycled chlorine and solvents, bore a label clearly marked "poison." Below that label was another with the words "United States of America" and the AID emblem, a pastel drawing of two clasped hands symbolizing good will and cooperation between the United States and other countries. How ironic, said Rossi, to find the "poison" label juxtaposed with the agency's emblem, which is placed on commodities whose sale is financed by the U.S. government.

Through a series of long-distance telephone calls, Rossi, then stationed with AID in Kenya, tracked the shipment that arrived in Zimbabwe back to New York and to Jack and Charlie Colbert. While working on the case, Rossi soon learned that he wasn't the only one on the Colberts' trail: The brothers were on a lot of wanted lists.

State and federal environmental agencies all along the U.S. eastern seaboard, Rossi learned, had either investigations or open files on the Colberts. "Their concern was: 'How do we stop these people?'" said Rossi. "'How do we get them off the streets?'"

The trail from Zimbabwe led Rossi to the Colberts' headquarters: a three-story warehouse just north of New York City. Here, in the suburb of Mount Vernon, the Colberts operated through three companies: SCI Equipment and Technology, the Mount Vernon Trade Group, and Signo Trading. Rossi paid a visit to the offices of Signo Trading and SCI Equipment, where he tried to question the Colberts, but to no avail. "They refused to talk to me at that point in time," Rossi recalled matter-of-factly. "They wouldn't even open the door."

According to Jack Colbert, this wasn't evasiveness. Jack said he wasn't even at his office when Rossi came by, and charged that the government—or at least part of it—was out to get them. "At the time that this Mr. Rossi came to see us, we were getting all sorts of government harassment. . . . It comes from the fact we had chemicals that we got from the government—and that was from one side of the government—and they said they were fine. Another side of the government, the environmental protection people, said they weren't."

Colbert warehouse in Newark, New Jersey.

Photo: Andrew Porterfield

All in all, it took four separate attempts by federal agents to serve the Colberts with the original grand jury subpoena in the case.[10] Al Rossi succeeded only after he decided to go up the fire escape at the back of the building, hoping that perhaps a back door would be open so that he and his agents could gain entry. "Fortunately for us," he recalled, "they had left the door open. I walked in and Jack Colbert was sitting there. . . . He got a little hostile with us, and said, 'What are you doing? You're trespassing!'"

The agents had been forewarned. In an earlier civil lawsuit, the plaintiffs' attorney felt compelled to inform the U.S. marshal that, when one of his employees tried to serve a subpoena on the Colberts, "he was physically ejected, and then they [threw] the papers out the door. Those people are dangerous."[11]

Rossi discovered that environmental protection agents on the East Coast were worried most about how the Colberts stored their toxic materials. "Their modus operandi," recalled Rossi, "was to sell off whatever they could . . . and what was left over they would leave in the warehouse. And they would abandon the warehouse."

The Colberts' string of toxic warehouses, often stuffed with explosive, flammable, and leaking chemicals, threatened the safety of local communities. In Mount Vernon on June 1, 1983, police and fire sirens sounded an alarm after a serious chemical spill of an insecticide forced neighborhood residents to evacuate houses and apartments on both sides of the warehouse. Thirty-one people, including police and fire fighters, were overcome by fumes and had to be rushed to the hospital.[12]

"Not only were [Colbert warehouses] right here on the border of New York City," Mount Vernon fire chief Henry Campbell recounted in amazement, "but we found that they were in Newark, that they were in Delaware, and that they had another plant in Mount Vernon at one time that we had cleaned out previous to this one. We don't know where they are." Authorities say the Colberts had warehouses in at least eleven other cities, including Baltimore, Trenton, Buffalo, and Greenville, South Carolina.

After a fire damaged the Colbert warehouse in Newark on April 11, 1983, authorities decided the only way to dispose of some of the chemicals safely was to destroy them.[13] The Colberts insisted, however, their operation was safe. "We never had a fire. We never had a chemical accident," Charlie proclaimed, blinking his eyes nervously.

"There was . . . only one fire," corrected Jack.

"And there were no tests that showed that any material was anything but virgin material," added Charlie. "Out of all the accusations."

There were, in fact, many accusations. Court files show that customers from Kuwait to India complained about the Colberts and their "virgin material."[14] But Jack Colbert remembered their dealings differently.

"Everybody was happy—people weren't complaining," he claimed. "The paint was paint, the solvents were solvents, the cans of lithium hydroxide were lithium hydroxide. There was no problem."

When asked about a lawsuit with a company in India, he replied, "India? . . . In India we had nothing to do with it." And what about the complaints from Kuwait?

"The material you're talking about in Kuwait," said Jack, "that came from Alchem-Tron, okay? They gave us invoices that said it was toluene diisocyanate, okay? We never checked it, we didn't know."

And the Chemplex case in Zimbabwe?

"All right," acknowledged Jack. "Well, Zimbabwe, of course, I don't know what happened because . . . I never knew the material was the wrong stuff; I would have never shipped it."

The Plutonium Connection

The Colberts' operations apparently did not stop with toxic waste and chemicals. By their own admission, they were also involved in another dangerous business: arms dealing. Jack Colbert said that he legally sold some munitions and had licenses from the U.S. State Department and other federal agencies to sell additional millions of dollars worth of weapons.

The FBI investigated reports that the Colberts had planned to sell ten pounds of plutonium to a Middle Eastern country, according to a government memo submitted by the prosecution. When asked whether he sold plutonium, perhaps the most toxic material known, Jack was less forthcoming:

"No. Well, I really don't prefer to talk about that, but if you want to find out I suggest you contact the United States government because they know the story on that very well."

U.S. officials tell a disturbing tale. "[The Colberts] offered to sell plutonium," according to former Assistant U.S. Attorney DeVita. "They got in touch with an individual with whom they had prior dealings in a variety of commodities, and he went to the FBI. The FBI had an undercover agent pose as a purchaser for a Middle Eastern country and negotiated to purchase from the Colbert brothers ten pounds of plutonium. . . . I believe it was $4 million a pound."

The Middle Eastern country the Colberts were willing to sell to was Libya. Although some federal investigators believe the brothers may have been only bluffing, DeVita said that on a January 1982 FBI wiretap, "Jack Colbert gave descriptions of the containers and the sensations of being in the room with plutonium which sounded very authentic."

And why would anybody want to buy plutonium on the black market?

"To make atomic weapons," said DeVita.

48

In 1986, Jack and Charlie Colbert were brought to trial as a result of Al Rossi's investigation of their shipment to Zimbabwe. They were each convicted of twenty-seven counts of conspiracy, mail fraud, wire fraud, making a false claim against the government, making false statements to the government, and one count each of obstructing justice. In July of that year, both brothers were sentenced to thirteen years.

Three toxic-waste transporters—Atlantic Coast Environmental Services, Resource Technology Services, and Alchem-Tron—were also charged with violating environmental laws.[15]

Al Rossi, for one, believes the Colberts might still be getting away with their toxic scam if the federal government had not inadvertently become involved because of the Zimbabwe waste deal.

Asked if it would be relatively easy to make the world a dumping ground for U.S. hazardous waste, Rossi had a quick reply. "Absolutely," he said. "To make the world a global dumping ground and to make a profit at the same time."

To date, clean-up of the Colbert brothers' chemical warehouses alone is estimated to have cost U.S. taxpayers in the tens of millions of dollars,[16] and other warehouses are still turning up. According to an EPA official, "There's potentially one in every state."

And the toxic waste the brothers shipped to Zimbabwe? No one is sure whether the chemicals simply evaporated or were dumped. One thing is certain: Zimbabwe has no facility for handling toxic waste.[17]

**Used chemical barrels line the streets
of a poor community near Tijuana.**

FOUR

Troubles on the Border

Not long after the Colbert brothers had left a toxic trail from Zimbabwe to Kuwait, U.S. officials were embroiled in another investigation much closer to home. For more than a year, investigators followed a Southern California waste broker and a Mexican truck driver through the shadowy world of hazardous-waste dumping along the U.S.-Mexico border.

Partners Raymond Franco and David Torres were the first people indicted on felony charges under a U.S. environmental law for smuggling waste from California to Mexico. They are unlikely to be the last. The Franco-Torres case provides a glimpse of a major law-enforcement problem, as well as what could amount to a toxic time bomb for those who live and work near illegal dumping grounds.

Each year, more than 130,000 trucks carrying lumber, scrap metal, and other supplies cruise down Interstate 5 and roll through the California-Mexico line at San Ysidro, one of the world's busiest border crossings. David Torres, who drove one of these trucks, allegedly hid barrels of toxic waste among his legal cargo. In fact, officials charge, Torres was able to smuggle scores of fifty-five-gallon drums of hazardous waste across the border before authorities caught up with him.

Torres was never stopped at the border, which is hardly

surprising. The chances of being detected are remote, according to U.S. Customs officials, who say they check not the cargo, but only the accompanying paperwork, which is easily doctored.[1]

Although U.S. Customs agents stepped up their spot checks for hazardous waste in 1990, they concede toxic shipments are not their highest priority; they are looking for arms, drugs, and illegal immigrants coming north from Mexico. The Customs Service also lacks the skilled personnel and on-site testing equipment necessary to check for toxic waste, which is difficult to detect. As one police officer observed, the trucks do not have "a placard on them saying they're hauling hazardous waste."[2]

Many law-enforcement officials agree, however, that the border traffic in toxic waste is on the rise. Faced with a dwindling number of landfills and soaring disposal costs that today run as high as $1,000 per barrel, some California firms find the temptation to dump waste across the border hard to resist. "Railroads told us they get shipments of waste down to Mexico every day," said one EPA official. "You just have to put a few drums on those cars and no one will know."[3] This view is shared by a Southern California deputy district attorney: unscrupulous companies and waste haulers, he said, consider Mexico "a big trash can."[4]

Strike Force Bust

Raymond Franco, according to investigators, was a street-wise entrepreneur, and the service he offered—helping businesses unload their hazardous waste—was much in demand. For at least two years, he ran Ray's Industrial Waste, a small operation out of El Toro and Burbank, two cities in the sprawling Los Angeles basin. On a typical day, Ray Franco could be found calling on firms that produce hazardous waste and explaining how, for a reduced rate, he could take toxic materials off their hands and dispose of them at a legal dump. One of those companies was Barmet Aluminum Corporation, a Torrance, California, firm that paints aluminum coils for wholesale distribution. In 1988, according to court records, Barmet officials made what seemed a good business deal: for about $12,000, Franco would dispose of fifty-seven barrels of their hazardous waste.

Although investigators estimated that to transport and dispose of Barmet's waste legally would have cost at least twice that amount, company officials said they believed they were hiring a reputable waste broker. But Barmet officials explained that the hazardous-waste facilities that were supposed to receive the company's toxic trash from Franco never did. "[What happened was] he defrauded us . . . several thousand dollars," said Barmet's Torrance division plant manager Jorge Eulloqui. "We reported the matter to the EPA."[5]

The EPA referred the report to an unusual team of investigators, the Los Angeles Environmental Crimes Strike Force, whose mission it is to nab toxic polluters. The strike force includes representatives from about ten agencies, including the Los Angeles District Attorney's Office, the Los Angeles County Health Department, and the California Highway Patrol, who work together as a kind of environmental SWAT team.

The special unit was set up to catch illegal dumpers in the act. In working undercover, conducting border roadblocks, and checking suspicious-looking trucks, it has met with some success. During the 1980s, the strike-force investigations led to the jailing of dozens of violators. In pursuing the tip on Franco, the strike force learned he had a criminal record and his business practices had aroused the suspicions of law-enforcement investigators and county health officials. So, in late October 1988, highway patrol investigators began tailing Franco to gather evidence of wrongdoing. It didn't take long.

"We put Mr. Franco under surveillance," said Los Angeles deputy district attorney William Carter, who prosecutes environmental crimes, "and the first day . . . we saw him engage in activity we believed to be suspicious." At one site, undercover investigators, one parked in an unmarked car and another perched on a neighboring rooftop, took hundreds of photographs of Franco loading drums of suspected hazardous waste into trucks, which he and others then concealed with cardboard, wood, and empty drums.

This surveillance of Franco also led the strike force to David Torres, who helped Franco load the trucks. Torres, a Mexican trucker who also owned a pottery factory in a working-class neighborhood of Tijuana, said he hauled material for Franco—which Torres could resell or use himself in Mexico—when his regular hauling business was slow. "I needed some money to survive," he explained in an interview.[6]

The plan was simple: Torres said he would pick up the drums of waste for Franco, who paid him $30 a barrel for his services. "I just go to the back of the building and pick up the drums and exit . . . with Franco," said Torres in halting English. "I don't know if I do something illegal."

Torres said he met Franco through Jack Rust, a seventy-three-year-old former chemical salesman. In 1986, Rust, while admitting no guilt, paid the EPA $10,000 for cleaning up what court documents called "corroded, distended and leaking" drums of hazardous waste that he allegedly abandoned at the border.[7] At that time, according to Torres, he was one of Rust's drivers. The Franco-Rust connection didn't end there: Rust said that Franco used him to get information and to enlist the services of Torres, who had access to other drivers.[8]

Rust, who now describes himself as an environmental

consultant to waste generators, thinks he knows why some companies turn to operators like Franco: the hazardous-waste disposal problem hits small businesses the hardest. Large firms have the money and the personnel to find creative solutions to the problem, but "mom and pop shops" simply do not. As a result, some small operators may find what Rust called "ingenious ways" to dispose of waste, such as hiring illegal dumpers. "A lot of my friends who would not listen to me have been in jail and have paid huge fines for not listening to me," Rust said.

One company that may wish it had listened to Rust is Barmet Aluminium. Rust said he was asked to bid for the job of taking away Barmet's hazardous waste, but company officials told him his fee was too high. In the end, it was Franco who won the contract.

Months of scrutinizing Franco's activities eventually led investigators to David Torres's run-down pottery factory in Tijuana. In February 1990, assisted by Mexican officials and clad in protective "moonsuits," investigators searched the factory, turning up barrels of chemicals that appeared to be the ones Torres and Franco had picked up from Barmet. Many of the drums bore bright yellow labels identifying them as hazardous waste. Filled with chemicals, solvents, and old paint considered poisonous and carcinogenic by the United States government, the barrels were found sitting just a few yards from an elementary school.

After more than a year of surveillance and search warrants, and with the assistance of the FBI, strike-force members successfully wrapped up their investigation of Franco and Torres. Officials said it was the first time in California that a law-enforcement surveillance team actually witnessed the attempted smuggling of hazardous waste into Mexico.

On May 9, 1990, Franco and Torres were indicted for felony violations of the Resource Conservation and Recovery Act (RCRA), a federal environmental law that forbids the transportation of hazardous material to an unlicensed facility, as well as the disposal of hazardous waste without a permit. Specifically, the two men were charged with conspiracy and the illegal transportation, disposal, and export of hazardous waste.[9]

If convicted, Franco faces a maximum sentence of thirty-two years in prison, and Torres, twenty-seven; both could be fined as much as $250,000 for each count. Raymond Franco's trial is pending. And David Torres? After the indictment, he chose not to return to the United States from Mexico, where he is believed to be a fugitive.

When county and federal officials issued the indictment, they also announced the creation of a new interagency Task Force on Environmental Prosecutions, with the FBI, EPA, U.S. Attorney's Office, California Highway Patrol, and state health department investigators all working together to solve environmental crimes.

A top priority of the task force, officials say, will be to track down the illegal export of hazardous waste from California to Mexico. Robert Brosio, an assistant U.S. attorney in Los Angeles, believes that the pursuit of waste traffickers is no less important than that of drug lords. Brosio stated: "We cannot be any less diligent in our efforts to bring to justice those who are smuggling American-produced hazardous waste out of the country."[10]

Illegal Dumping Close to Home

How many others like Franco and Torres are out there? No one really knows. In Southern California, despite increased efforts by authorities, only five highway patrol investigators are assigned to search for illegal waste leaving the state. The EPA, in charge of policing hazardous waste, is stretched even thinner: it has just one agent and two investigators to enforce the RCRA and several other statutes in four Western states, as well as the far-flung U.S. territories of the Pacific.

Southern California produced a little over half of the state's hazardous waste in 1988,[11] and prosecutors there say most of their cases involve illegal dumping within the state: drums abandoned in vacant lots, containers emptied into sewers, and barrels ditched along deserted stretches of California's highways. But law-enforcement officials agree that waste trafficking across the border is a troubling issue. "I wouldn't call it a trickle," said strike force member Carter, choosing his words carefully. "I wouldn't call it a flood. I think it's more of a steady stream."

During the late 1980s, reports of illicit dumping made headlines in Southern California newspapers, including the news that shipments of potentially harmful used printing ink and solvents were sent to a plant near Tijuana that local authorities said had no import license or waste-disposal permit. Ironically, the waste had come from the bearers of the news, Southern California newspapers, including the *San Diego Union, San Diego Tribune,* and *Orange County Register*. The papers had chosen to recycle their waste ink through a hauler whose services they believed were legal, safe, and inexpensive. The hauler, it turned out, was passing on the waste to the unlicensed Mexican facility.

The factory posed a hazard to workers and residents alike. One Southern California reporter investigating the incident discovered that a boiler, jerry-rigged from a used American railroad tank car, contained a potentially dangerous mixture of oil and ink used for making asphalt.[12] The car bore the label: "Not for flammable liquids," and stood next to a boiler charred by an accidental blaze that had occurred several weeks earlier. Living in a house only ten feet away from the boiler was sixty-three-year-old Vasilia

Cabrera de Lopez, who complained that fumes from the factory and a nearby battery plant had made her ill.

The waste ink fiasco was uncovered by California legislator Steve Peace and his aides, who studied hundreds of state-mandated shipping forms, known as hazardous waste manifests, and federally-required export notification documents. While searching through the piles of paperwork, they also discovered that the newspaper companies shipping waste ink to Mexico had not filed the papers required by the EPA.[13]

Illegal Dumping Across the Border

Along with the stream of hazardous waste entering the country from north of the border, Mexico is being polluted by U.S.-owned companies inside the country. About 1,800 such factories, known as *maquiladoras* or "twin plants," produce everything from batteries and toys to pesticides and food products. They take advantage of Mexico's low wages to manufacture goods that are later shipped back to the United States for sale.

Waste from the *maquiladoras* is supposed to be shipped back, too. But despite a 1983 agreement between the two countries on this issue, authorities fear that untreated hazardous waste is being widely dumped inside Mexico. Used chemical drums from *maquiladoras,* which often contain poisonous residues, are a common sight in the poor Mexican neighborhoods known as *colonias.* In a home in Colonia Anapra, outside Juarez, one reporter found a family who stored their drinking water in a used U.S. chemical drum labeled: "CORROSIVE. Sodium Hydroxide Liquefied Formula 291. Precautions—causes severe burns to skin and eyes. Swallowing results in severe damage to mucous membrane and deep tissues."[14]

Under the terms of the 1983 agreement, and a later annex, companies are supposed to contact the EPA when they bring their waste back into the United States. But EPA officials said they received only ten such notifications for California and Arizona in all of 1989.[15] Given this low number, even the staunchest industry advocates concede that illegal dumping is a problem. As Dan Pegg, chair of the Border Trade Alliance, told one reporter, "There's no question that hazardous waste ends up where it's not supposed to be."[16]

One place pollution from *maquiladoras* ends up, say environmentalists, is the New River, which flows from Mexico's Baja California peninsula into California's Imperial Valley. U.S. health experts consider the New River one of the most polluted waterways in the world; scientists have tested the river and found it contains as many as 100 toxic and cancer-causing chemicals. California health officials also say the river is loaded with harmful viruses that could lead to a major public health catastrophe. In late 1986, Imperial County's then-public health director told the CBS television program "60 Minutes," "I think this river is a disaster waiting to happen."[17]

Unfortunately, the New River is not the only potential disaster along the U.S.-Mexican border. In Ejido Chilpancingo, a poor community near Tijuana, open drainage ditches are contaminated by sewage and toxic discharges from nearby Mexican factories and U.S.-owned *maquiladoras*. Local residents, who live in tiny houses fashioned from cement blocks, wood, nails, and wire, say the water makes their families extremely ill.

One mother, Maria de Consuelo Montoya, reported that her children's skin was covered with blisters, their mouths infested with sores, and their hair had fallen out from drinking the polluted water near her home. Dr. Juan Manuel Sánchez León, who examined Montoya's children, agreed

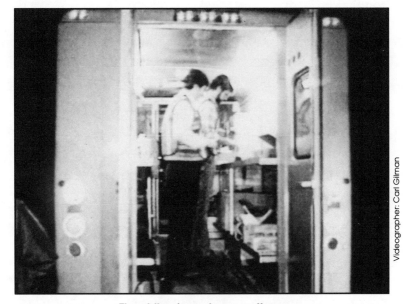

Videographer: Carl Gilman

**The strike-force team gathers
evidence to catch toxic polluters.**

that "bad water" from the factories was partly to blame for their ailments.[18]

In tests conducted in early 1990, the well water the family uses for washing was found to contain high levels of cadmium, a carcinogen linked to liver and kidney disease. Soil samples from the area[19] also revealed elevated levels of heavy metals, including lead, which is known, at even low amounts, to cause damage to the central nervous systems of children.

The sick children's grandmother, Rosa Devora, has a message for the owners of the *maquiladoras*: "They should not throw any more contaminated water our way," she said calmly in Spanish. "We are humble people. . . . They should treat us with more understanding."

The *maquiladoras* provide many jobs for Mexicans who might otherwise be unemployed, but people like Devora have also argued that Mexico and the United States need to confront the health threat posed by these thriving factories. "It is good that they provide jobs for our sons and daughters, for us and our welfare," she said. "But if they give us jobs on one hand, but on the other hand we are sick," she gestures sadly. "That is the problem."

And it is a problem that many citizens and government officials argue is not being adequately addressed. The Mexican equivalent of the EPA is badly understaffed and underfunded. As a result, the agency, known as the Secretaría de Desarrollo Urbano y Ecología (SEDUE), has found it difficult to enforce environmental regulations. SEDUE officials, however, have been cooperating with their U.S. counterparts in an effort to clamp down on illegal dumpers. "Our job gets more difficult, complex and larger every day," said Jorge Escobar Martínez, chief of SEDUE's Department of Prevention and Control of Environmental Contamination.[20]

One thing is clear: California's waste-disposal problems are only getting worse, and this is bad news for Mexico.

Some strike-force members, for example, think of Los Angeles County as a toxic pressure cooker where thousands of businesses and government agencies classified as hazardous-waste producers are facing strict new regulations.

No one knows how many Southern California businesses will be tempted to turn a blind eye to suspiciously cheap waste haulers who promise to get rid of their waste. Unless businesses find alternatives, some law-enforcement officials fear the worst. One strike-force investigator summed it up this way: "The bottom line is money . . . the almighty buck."

Strike-force officials also warn that waste dumped in Mexico may affect U.S. residents as well as Mexicans. "The reality is, it's going to come back to us in some way or other—air pollution, water pollution. It might come back to us in the products we buy," said environmental crimes prosecutor Carter. "It's poison you're sending down there. It will kill these people."

"The intentional disposal of hazardous waste," Carter concluded, "is no less harmful than someone planning to bomb someone. This bomb may not go off in a minute or a second or an hour. It may go off twenty years from now."

Workers sort U.S. batteries in Taiwan.

FIVE

Loopholes in the Law

Inside a securely-guarded factory in the farm country of south central Brazil, workers toil in a steamy heat made more intense by the inferno of the plant's furnace. Sweating heavily, some remove their tight-fitting protective masks as they go about the poorly paid task of melting down lead waste into metal bricks. Suddenly, a thermal shock in the furnace causes a small explosion. Choking smoke fills the factory and lead fumes curl into every corner. As the smoke clears and the startled workers survey the scene, they find dozens of birds that nest in the factory's rafters have fallen dead on the floor. Like the canaries that coal miners once used to test for lethal gas, the birds provide grim proof of the plant's toxic dangers.

This scene, at the Tonolli lead recycling plant outside the industrial city of Sao Paulo, was recounted by the factory's former laborers, who didn't need birds dropping dead at their feet to know they faced health dangers on the job. They complained of a long list of symptoms, all related to chronic lead poisoning: headaches, dizziness, stomach cramps, nausea, kidney pains, aching in their arms, legs, and joints, and an overpowering weakness.

They protested, as well, the powerful side-effects of the "detox" drug the factory gave them to counteract their lead exposure. Some chose not to take the toxic medication, which can cause nausea and kidney damage, stockpiling the pills at home instead. Meanwhile, local health officials voiced concern that the factory was threatening the surrounding community after hearing reports that the company had buried lead residues in large pits that might contaminate local water supplies.[1]

Much of the lead causing these problems comes to Brazil from overseas, particularly from Canada and the United States.[2] Often it arrives in the form of used batteries—the same as those found under the hood of virtually every car. Used batteries contain poisonous lead and acid, but because of an exemption in U.S. environmental regulations, batteries that are still intact and exported for recycling are not classified as hazardous waste.[3]

This loophole means the U.S. EPA does not have to notify the importing country about the hazardous nature of any shipment of whole batteries destined for recycling. It also means U.S. batteries that consumers dutifully return for recycling may wind up overseas, poisoning both people and the environment.

According to industry estimates, at least 70 million automobile batteries will be discarded in the United States each year during the 1990s—a figure that translates to roughly 70 million gallons of sulfuric acid and more than a billion pounds of lead. Although the United States has one of the world's safest and most sophisticated systems for recycling its used batteries, anywhere from 10 to 20 percent of them end up dumped unceremoniously by the side of a road, thrown away with the regular garbage or just left in a garage and forgotten.[4] Of the 80 percent or more that are recycled, a substantial number are sent overseas, where they are smashed apart, melted down in lead smelters like those in Brazil and poured into ingots for reuse.

In addition to Brazil, millions of used U.S. batteries were exported in 1989 to lead smelters in Mexico, Japan, Canada,

India, Venezuela, China, South Korea, South Africa, and Taiwan.

"We have no control over those [exported] batteries," said Wendy Grieder, an international activities specialist for the EPA's Office of International Activities. Unless they are already cracked and leaking acid, she explained, "under present regulation, we have no control over those batteries that are whole, that are simply sitting [overseas] causing problems."

The EPA is well aware, however, of the problems that used batteries can cause. Another Tonolli battery-recycling plant in Pennsylvania, for example, owned by the same Italian company that once operated the Brazil plant, caused major environmental problems for the agency after the plant went bankrupt in 1985. The EPA announced in 1987 that the factory was a "Superfund" site, which means it had been classified as one of the most dangerous hazardous-waste sites in the country.[5] As of 1990, the average cost to clean up a Superfund site was $25 million.[6]

Caution: Lead Recycling

The lead in a car battery may be successfully recycled and still cause severe damage because lead is one of the most inherently dangerous substances on earth. It has been considered a significant health hazard for thousands of years, dating back to ancient Rome, when Pliny the Elder wrote about the element's toxicity. More, in fact, has been written about lead as a toxin than any other substance. And the potential dangerous effects on lead-smelter workers, like those at Tonolli, have been known for hundreds of years.

Among other things, lead can harm the brain, nerves, kidneys, and reproductive system. Typical symptoms of major exposure to lead are high blood pressure, gout, inflamed kidneys, tremors, lassitude, and a leakage of the kidneys that causes protein and sugar to appear in the urine. Because lead assaults the central nervous system, people with lead poisoning become weak, and their thinking is impaired. Some lead workers may develop kidney cancers, depending on their exposure to other hazards. Lead can also damage the reproductive systems of both men and women, resulting in impotence, reduced sperm counts, and lower fertility.[7]

The effects of lead on children are even more severe, including kidney disease, gouty arthritis, mental retardation, and psychological disturbances. Although adults eliminate most of the lead they breathe in, children absorb around 40 percent of it.[8] Even relatively small amounts of lead appear to lower children's IQs, and in 1990, a major study by U.S. researchers showed the effects may be permanent, resulting in high dropout rates from school, reading disabilities, behavior problems, and the loss of job opportunities and earnings.

In Brazil, smelter workers at the Tonolli factory worry about the impact of lead on their own future. Medical tests by local public-health officials between 1987 and 1989, for example, showed that present and former workers from Tonolli and another lead smelter called FAE had serious lead contamination problems: sixty-two percent of the FAE workers and twenty-five of twenty-nine Tonolli workers tested above acceptable U.S. levels for lead exposure.[10]

Some workers at Tonolli had blood-lead levels two to three times the U.S. standard, putting them at high risk of developing nerve damage. These and other workers claimed the company had forced them to take a strong "detox" pill containing EDTA, sometimes as many as six a day, to keep their lead levels within an acceptable range. A Brazilian state occupational health advisory warned that this medication should only be taken intravenously, because "oral administration of [this drug] is not advised and not done anywhere in the world."

In 1988, a local public prosecutor's office in Brazil launched a police inquiry to determine whether it could file criminal charges against Tonolli for abusing its workers' health by exposing them to lead. So far, no charges have resulted.[11]

That same year, a local paper reported that the state environmental agency responsible for the Tonolli factory fined the plant after five cows died on a nearby farm and lead was found in the farm's pond, spring, and pastures.[12] Brazilian authorities, however, have been slow to crack down on that country's lead imports. "If we block those imports, the company will close," said one state environmental official. "And we're not in the business of closing down companies." A Brazilian law passed in 1990 has sought to improve the oversight of all waste imports, particularly scrap metals, by placing greater responsibility on exporters to screen their waste.[13]

Dangerous conditions like those at Tonolli exist in battery-recycling plants throughout the underdeveloped world. Another battery-reprocessing plant on a hill overlooking a poor neighborhood near Tijuana, Mexico, regularly processes used American batteries, but provides its workers with little protective clothing. Samples of surface water near the plant showed levels of chromium, copper, and lead exceeding Mexican standards for the discharge of industrial waste water. Soil samples outside the plant showed amounts of lead that far exceeded safety levels for hazardous waste in California, just a few hundred yards away.[14]

In the United States, on the other hand, the high level of toxicity associated with lead smelting has led to stringent safety and environmental regulations. But that does workers overseas little good: America's stiffer rules helped cut the number of U.S. lead smelters in half from 1980 to 1986,[15] further spurring export of the country's used batteries to places where labor is cheaper and the laws more lenient.

Taiwan Cracks Down

Dr. Jung-Der Wang is a cautious, religious man, hardly the type to cause a firestorm of controversy in his native Taiwan. But Dr. Wang, who studied at Harvard, was one of a handful of doctors specializing in the treatment of environmental health problems when he returned to the overcrowded island in 1982. His research there has produced some alarming information.

In 1987, a forty-two-year-old worker approached Wang, complaining of faintness and weakness in his limbs. Wang soon discovered that the lead level in the worker's blood measured more than twice the U.S. occupational limit. The likely source of the contamination was his patient's workplace: the sprawling Acme lead smelter near the northern port city of Keelung, a major importer of used batteries from the United States and Japan.

Acme is one of two smelters on Taiwan that had become a magnet for used batteries and scrap lead from overseas. During the 1980s, Taiwan became Asia's leading importer of U.S. lead scrap, with regular shipments from ports such as Long Beach, Oakland, and Seattle. Until Dr. Wang, however, no one seemed to have questioned what effects the massive lead imports might be having on the population of Taiwan, an island nation half the size of Panama and packed with 20 million people.

When Acme managers were reluctant to cooperate, Dr. Wang obtained the help of government officials to examine other workers at the plant. The results were as he feared: of sixty-four workers, thirty-one had lead poisoning, some with blood lead levels almost three times the U.S. limit. Worried that contamination had reached the surrounding community, Wang next turned his attention to a kindergarten located near the plant.

"The reason I wanted to do a study is because if there is damage, then I have to warn the kindergarten so that they can move," Wang explained. The effects on children, Wang knew, would be particularly serious because of their greater susceptibility to the effects of toxic materials. There was indeed reason for concern: of the thirty-six children he examined, twenty-two had elevated lead levels. The children were immediately moved to another school.

Before Wang could publish his study, word of his findings was leaked to the press in early 1990 and provoked a wave of public concern. The environment had recently become big news on Taiwan: a 1989 blue ribbon panel warned that the island risked becoming "a poisonous garbage dump."[16] Now, with Dr. Wang's findings, the image of children being poisoned by careless business people enraged citizens across the island. Adding to the controversy, a newspaper revealed that thousands of tons of waste from Acme had been dumped onto an open field nearby and were threatening water supplies for the surrounding community. Before emotions calmed, protesters marched on Taiwan's other lead smelter that imports U.S. batteries, 200 miles away, demonstrating and smashing windows.[17]

Among the experts studying the situation on Taiwan was Michael Rabinowitz, an American geochemist from Harvard University working at the invitation of the Taiwan government. Rabinowitz, a specialist on the effects of lead exposure on children, confirmed the dangers faced by workers when, in April 1990, he donned a hard hat and toured Thai Ping, the largest lead smelter in Asia.

"From what the factory manager tells me, the blood-lead testing of these workers [at Thai Ping] shows that they have blood-lead levels in the range of 40 to 80 micrograms per deciliter," he explained. "That puts them at high risk for developing kidney problems and also nerve problems."

"I can see a lot of these [Thai Ping] batteries are from Seattle and other [U.S.] ports," he added, shouting over the noise of machinery in the background. Although the lead from American cars is giving people in Taiwan jobs, "as a result of the way they're doing their jobs, many of them will probably get sick."

Rabinowitz said that his studies of the lead levels in schoolchildren's teeth in the agricultural area near the Thai Ping plant revealed they were about twice as high as those for the average city-dwelling child in the capital city of Taipei. Children who live or go to school near smelters become exposed to lead when it escapes the grounds by being carried home on the clothing of the workers or is carried away in the air or water. "These children," warned Rabinowitz, "can be expected to have impaired intelligence, slower physical growth and some behavioral disorders—trouble paying attention, hyperactivity."

As evidence, he cited Dr. Wang's study of the children living next door to the Acme smelting plant. "These children seem to have lower IQs according to how much lead they're exposed to," explained Rabinowitz. "And what's very interesting to me is: the children who are most affected are the children who come from better social classes. So the children you expect to do really well, the children you expect to develop into the leaders for tomorrow, these are the children who are most sensitive. They are most affected by the lead." He said he was hopeful the children could still recover if their exposure stops.

Based on his research, Dr. Wang recommended that Taiwan's government stop the import of batteries. "I consider IQ one of the most important things," Wang explained earnestly. "The neuro–behavior system is very important because these children are the future people of . . . Taiwan. We have a lack of resources, other than people. We have to take care of our people."

Contamination from the lead smelters was so extensive that by 1990, Eugene Chien, head of Taiwan's newly-formed Environmental Protection Administration (EPA), decided to end all battery imports. In addition, Chien proposed to upgrade the island's smelters so they could begin to reprocess Taiwan's own batteries safely. The lesson he's learned, said Chien, is "don't import from the United States—it causes too many problems for us."

Chien is a savvy forty-four-year-old administrator with a Ph.D. in aeronautics and astronautics from New York University. Since assuming the top position at Taiwan's EPA when it was established in 1988, Chien has energetically tackled the island's massive environmental problems one by one.

Used car batteries are not the only U.S. waste export that cause him concern. "We have a very special industry called the scrap metal industry," explained Chien. "Definitely, [this] business is no good for our country's health. . . . It causes so much of a problem with air pollution, water pollution, and very serious damage to our rivers."

On the surface, the export of scrap metal may seem innocuous enough, but shipments arriving at the docks of Kaohsiung harbor, one of the world's busiest ports, include everything from transformers filled with PCB-contaminated oil to old electrical equipment containing asbestos. Much of the scrap metal is bound for processing in junkyards that provide no masks or other equipment to protect the workers against asbestos, PCBs or metal dust. The processing itself also includes dangerous and illegal practices such as burning copper cable, a method that releases poisonous fumes.

The industry's pollution has spread to wherever the scrap dealers have located. "In a certain part of Taiwan, we

**Joe Chen supervises work at his
Guo Fu scrapyard in China.**

have a small river," Chien explained. "The whole river is polluted by the scrap metal industry. And in some sense, you can call it the 'Death River' because it's no good for fish life."

Accordingly, Chien has decided to phase out the scrap-metal industry on Taiwan by 1993, removing in the process a $100 million annual business with powerful political connections. Nearly seventy percent of the scrap waste comes from the United States.[18] It has yielded copper, aluminum, platinum—even gold—and consisted of everything from computers and telephone systems to appliance motors, transmissions, air compressors, condensers, capacitors, and electrical wires.

A country like Taiwan, said Chien, simply does not have the resources to control this complex waste or clean up the resulting contamination. "As far as the scrap metal is concerned, you can see the tremendous amounts of waste there piled just like a mountain," Chien said, referring to a scrap processing zone in Taiwan. The scrap dealers, he charged, "take all the money out and they leave the waste to the government. So that's why we want to phase out all their programs. There are illegal and irresponsible businessmen."

"The King of Scrap"

One who takes issue with Chien and his EPA is Joe Chen, who some call "The King of Scrap." A Chinese-American based in California who is originally from Taiwan, Chen has been one of the biggest suppliers of non-ferrous scrap metal to Taiwan. From the headquarters of his Tung Tai Trading Corporation located just south of San Francisco, Chen is constantly on the move, traveling around the United States to his scrap suppliers, then taking the fourteen-hour flight to Asia to supervise the delivery of his scrap metal for recovery. Not surprisingly, Chien's action in 1990 to shut down all scrap metal imports into Taiwan by 1993 angered Joe Chen.

"It's very unfair," argued Chen, a genial man whose dark hair is streaked with gray. "You cannot blame all these buyers, all these people here [in Taiwan], because the government itself has got to be blamed. I don't think they had a good plan when they set up this [scrap processing] yard. I don't think they [the EPA] can name any place that disposes of garbage or hazardous waste."

Chen argued that the government on Taiwan has made money for years from fees on the scrap imports without establishing a safe disposal system for the remaining waste. Taiwan's booming scrap-metal industry, he said, should not suffer because Taiwan's EPA later determined there was a problem. The EPA should at least try to provide the scrap dealers with guidance as to what kind of materials could be imported.

Chen suspected that the EPA was merely reacting to complaints from people who lived near the scrap-processing zones or had demonstrated against other pollution problems. He also accused the environmental agency of corruption by letting hazardous materials continue to enter the country after local inspectors received payoffs.

For Chen, the scrap metal business is a form of alchemy, transforming "junk" from the United States into gold for Taiwan. He charged the EPA and the government were forgetting that, without a source of natural resources on Taiwan, the scrap industry had been a key source of the island's amazing prosperity. "For the last thirty years, the reason why Taiwan is so rich . . . is by buying all this kind of junk—this scrap metal, the low-grade stuff from all over. This makes Taiwan what Taiwan is today," boasted Chen. "I'm sure of that."

Joe Chen's defense of the scrap industry did not prevent

him from acknowledging the real environmental problems the scrap business has created in Taiwan. He conceded that the industry has profited at the expense of the island's rivers, air and soil. Much of the imported scrap contains hazardous materials like PCBs and asbestos, Chen believed. And, in the Ta Fa Industrial Zone in the southern part of Taiwan, claimed Chen, "ninety percent of the buyers burn . . . wire in the open sky," creating dangerously toxic fumes.

Outlaw waste dealers burn the scrap wire illegally in fields to retrieve the copper inside, at the same time releasing dioxins from the burned plastic wire coverings. "They burn it at midnight or in the morning," said the EPA's Chien of the scrap dealers. "You just need gasoline and a match. . . . It is just like hide-and-seek how they play a game with our enforcement officers." Students in schools near burning scrap, he said, have had to wear masks in the classroom to cope with the pollution. In some classes, children wrote with one hand while holding a handkerchief over their faces.

When Chien began in 1990 to phase out Taiwan's scrap trade, he met with reluctant U.S. officials to ask for their help in stopping export licenses for scrap companies in the U.S. "To our surprise," Chien remembered, "the U.S. government says that because of free trade . . . [Taiwan must] open the door for scrap metal."

Wendy Grieder of the U.S. EPA confirmed that tons of U.S. scrap metal continue to be exempt from U.S. environmental controls at the request of the Office of Management and Budget, Commerce Department, U.S. Trade Representative's Office, and Council of Economic Advisers. "I don't know that we're urging [scrap metal exports], but we're not hindering," said Grieder. Given a trade gap with Taiwan that is two to one against the United States, there are strong economic incentives for the government to keep scrap-metal exports flowing to that Asian country.

Strict U.S. rules for the domestic handling and disposal

of low-grade scrap have provided operators like Joe Chen further incentive to ship their scrap overseas for processing.[19] And now the crackdown on Taiwan is forcing Chen and others out of that locale, as well. But the resourceful Chen is undaunted. He has simply opted to move his scrap business to another country, to a place with fewer regulations and cheaper labor: the People's Republic of China.

The Move to the Mainland

In moving to the Chinese mainland, Joe Chen had a head start over his fellow scrap dealers on Taiwan. As a U.S. citizen, Chen is not bound by the trade restrictions imposed by Taiwan on trade with the mainland. By 1990, he had established a scrap-metal processing business near Zhuhai City in subtropical Southern China. Chen soon was shipping nearly 150 truckloads of scrap every month—ninety-five percent them from the United States—to this business, the Guo Fu smeltery located on the fifteen-acre site of a failed aluminum factory amid fields of rice, fruits, and livestock.

"Have a Nice Day" license-plate holders from California, used batteries, electrical motors, copper wire, even used IBM computers have all found their way in green truck containers from San Francisco to China. At the Guo Fu plant, workers are paid about two dollars a day to process this scrap into pure copper, aluminum and stainless steel.[20] Chen admitted that nickel cadmium batteries, transformers, and capacitors with PCBs from the United States had all been shipped to the plant, which he helps run in a joint venture with the Chinese government.

China, said Chen, was interested only in the foreign exchange that he could provide. "They supply the manpower and we supply the materials," he explained. "And they process the materials for us. We are requested to ship all the finished products out of this country."

"Right now I got the feeling the government [in China] only cares about the money," Chen said matter-of-factly. "I don't think they realize the problem yet."

Chen pointed to what may become a future problem for mainland China—a wide and deep ravine running hundreds of yards along the back of the Guo Fu factory. Water at the bottom of the grassy ravine drained into a pond where hundreds of ducks are raised for market.

"Now you can see from there, all the way around to back there, we are going to use this as the Tung Tai garbage dumping place," Chen said proudly, sweeping his arm across the ravine. He explained that he planned to dump the unusable plastic and hazardous materials such as oil from transmissions, air compressors, and transformers into the ravine. It would take approximately three years to fill up, he said, conceding that he would, in essence, be creating a hazardous-waste dump.

Although Chen insisted he was paying attention to potential environmental and occupational problems, his history of fines and lawsuits is not encouraging. On a San Francisco pier in 1986, a forklift collided with one of three Westinghouse transformers he had been storing for shipment to Taiwan. PCB oil poured out onto the warehouse floor at the pier, spurring calls to fire and health department officials and the mayor's office. The U.S. EPA ended up citing him for violations of the Toxic Substances Control Act[21] that, Chen admitted, cost him "a lot of money" and nearly resulted in a criminal charge.

Chen also had problems with the U.S. Coast Guard over shipments of used batteries to Taiwan—so much, that he eventually decided to get out of battery-waste exports. He admitted that one of the reasons the Coast Guard fined him was his improper labeling of the batteries as "lead scrap"

Videographer: Norman Lloyd

**Joe Chen at the Guo Fu
dumpsite in China.**

and "electronic components." "I did instruct the supplier to put a label for hazardous waste on the . . . batteries," said Chen. "But you know that not all the suppliers will stick by the rules."

Still, Chen believes that he could be trusted to avoid these problems in China. "I inform all my people . . . to watch all these kinds of [hazardous] materials. If they have any question about the materials, they should make a report and let me know right away," Chen said.

But Joe Chen's assurances prompt as many questions as they provide answers. At one moment, he insisted he would handle his waste in China carefully "because I learned all the lessons in the United States." At the next, he said the government of China would be tempted to cut corners, "to look the other way," when it came to environmental and occupational safety. Then he admitted that his workers at Guo Fu "tried to burn wire in the open sky because it seems like a big area, but we found out that we still got a lot of smoke and we don't like to make a big issue in the area." Chen stopped the burning to avoid further controversy.

Joe Chen's activities are not the only sign that the global dumping ground is moving to the Asian mainland—and that China is now interested in taking up where Taiwan is struggling to leave off. By 1989, China's imports of U.S. lead scrap were three times greater than Taiwan's, causing China to replace Taiwan as Asia's leading importer of lead scrap from the United States.[22] In 1990, the U.S. EPA disclosed a proposal from a U.S. company to operate a "hazardous waste reduction plant" on China's Hainan Island and a "hazardous waste plant" in Fujian Province that would process "toxic and non-toxic chemical materials and wastes."[23]

Eugene Chien at the Taiwan EPA believes the burden of a solution should fall to the source of the waste. "The problem actually relies on the United States because if they cannot export hazardous waste to the underdeveloped countries, then this problem is all solved," reasoned Chien. "If we [Taiwan] say no, and the U.S. continues to export the hazardous waste out, then they [waste traders] can ship from Taiwan to mainland China to Thailand to the Philippines."

Joe Chen, meanwhile, remains an aggressive businessman in search of a market. "It's thousands of tons every year in the United States, and the United States [has] got to find a place to dispose of it," he said. "So businessmen like me try to find a place to continue doing this kind of stuff."

Chen smiled when he considered the quiet, grassy ravine behind his mainland plant—just one more location on the growing list of dumpsites for the international trade in hazardous waste. "That's the perfect place," said Chen. With the support of the government, Chen opened up three plants on the mainland in 1990.

"People are talking about Malaysia, or Singapore, or Pakistan or India or the Philippines," Chen recounted. New energy came into his voice as he considered his prospects. "Yes, a lot of people try to find a place to get rid of their stuff. And I think the best place right now is China."

A rhyme protesting the import of hazardous waste from
West to East Germany: "The border keeps out people
but it doesn't keep out the poison."

SIX

The First World: A Toxic Gold Mine

A cres of smoking waste stretch out on all sides; trucks and bulldozers roar as they crisscross the vast dumpsite, shoving shapeless mounds of refuse from one spot to another. Piles of muck tower over the surrounding landscape of pastures and woods, in bizarre contrast to the medieval spires rising from the idyllic German town a few miles away. This is Schoenberg, which means "beautiful mountain," the site of Europe's largest waste dump. From time to time, fires have broken out at the dump, followed by occasional rains of ash on the nearby town.

Since the late 1980s, the world's attention has been drawn to the dumping of wastes in the Third World. But there is an even larger and equally troubling trade in toxics flowing from one industrialized nation to another. The United States, for example, ships waste to Canada; Western Europe, to Great Britain and the new democracies of Eastern Europe. Canada and the United Kingdom in particular have built flourishing businesses to take in unwanted materials from other nations.

Far from apologizing for what they do, executives of Canadian and British disposal companies promote themselves as environmental good citizens. As long as there are

PCBs in the world, they say, high-temperature incinerators are needed to destroy them.

But there is another side to this story. Too often, the First World trade in wastes has not turned out to be the clean solution to the waste problem that its managers and investors advertise. Instead, high-technology plants have attracted controversy, with reports of leaks, accidents, rashes of illnesses among neighboring families, and injuries to workers. Furthermore, just as the waste stream seeks out the path of least resistance *between* nations, so it does *within* nations. Disposal facilities are usually located in areas where people need jobs—and where people too often are either unaware of the problems such facilities can cause or not influential enough to keep them out. The First World trade in hazardous wastes also illustrates what happens when one industrialized nation has lower environmental standards than another, and it shows that even among technologically advanced nations, dumping can put human beings and the environment at grave risk.

Wales: The Vanishing Ducks

The Caldicotts' ducks disappeared without a trace. Ken Caldicott, a chemist, and his wife, Shirley, live with their two young sons in a valley in South Wales, their home situated within 30 yards of Rechem, Ltd.'s sixteen-year-old toxic-waste incinerator. They used to give away their ducks' eggs to neighbors who, like the Caldicotts, said they suffered frequent respiratory problems and other ailments that they blamed on the fumes issuing from Rechem's smokestack. The couple, who worried about the health effects of incinerator discharges on their children, videotaped the clouds of smoke drifting across their back garden, which, they said, sometimes happened three or four times a day.

Later on, tests of the duck eggs for PCBs by local authorities confirmed the family's fears; they showed contamination of up to 600 parts per billion, far in excess of internationally recognized standards for safety.[1] Alarmed that the PCBs had entered their food chain, the Caldicotts and other residents of the small town of Pontypool called for an investigation of the company. Then, one Sunday, while environment officials were still sampling eggs, the ducks simply vanished. "We previously lost odd ducks to mink," said Shirley Caldicott, "but six at once is very strange. We had them for just over a year since they were chicks and there is no trace, not a feather." Her husband called the disappearance "sinister." Because Shirley had gathered four eggs earlier in the week, she was able to provide samples for further tests, but the ducks never returned.[2]

The Caldicotts and their neighbors worry that they and other residents have become victims of the United Kingdom's flourishing waste-disposal business. In Margaret Thatcher's version of "can-do" capitalism, imported waste increased 20-fold between 1981 and 1989.[3] With environmental regulations weaker than those of most developed nations, Great Britain has provided an important outlet for the flow of West European toxic waste, and is exceeded only by East Germany in the amount it accepts. In a marked departure from First World safety standards, Britain allows hazardous waste to be mixed with ordinary rubbish and dumped into landfills. Costs are low, sometimes as little as £2 per ton. The United Kingdom takes in waste from at least sixteen countries, including fly ash from Switzerland, contaminated soil from Denmark and acid wastes from West Germany.

In 1986/87, Britain imported 53,000 tons of waste deemed "special," up from just 5,000 tons three years before.[4] "Special wastes" are substances that would be likely to cause serious damage to human tissue after an exposure of fifteen minutes or less, or after ingestion of more than a teaspoon.

A multiplicity of waste-disposal sites and decentralized regulations have created a haphazard system of enforcement in the United Kingdom, a situation described as "disgraceful" by the chair of a parliamentary committee investigating the issue. As recently as 1986, waste producers did not have to notify local Waste Disposal Authorities (WDAs) of the amount of hazardous wastes they were shipping. The seventy-nine WDAs were supposed to draft their own procedures for monitoring toxic substances, but by 1989, 75 percent of them had yet to do so.[5] Some local authorities kept their records in shoe boxes.

By the end of 1988, when operating landfills in the United States had declined to 325, Britain had 4,156 licensed landfills—an astonishing number for a small nation.[6]

Under this fragmented system, some waste shipments slipped easily through the cracks; others were discovered to be falsely labeled, with responsibility difficult to trace through a confusing network of brokers and transfer stations. Commented one environmental official: "It is an unlucky malefactor whose sins find him out. . . . Illicit loads will only be discovered when the landfill erupts in brown fumes." A 1988 investigation by the London *Observer* concluded that controls were "scandalously lax" and that the situation was getting worse.[7]

Environmentalists say that the waste trade succeeds because of government complicity with industry, and that, by creating more disposal sites, it discourages the development of cleaner technologies and waste reduction.

For years, the Caldicotts and other residents of Pontypool have complained bitterly of the consequences of the emissions from the nearby Rechem incinerator. Because the huge incinerator is situated at the bottom of a valley, the effluent fumes reach many townspeople. Its enormous smokestack ringed with stairways dominates the town, its tidy homes, gardens, and pastures. Residents complain of nausea, sore eyes, mouth ulcers, breathing problems, and anxiety over the plant's emissions.

In sharp contrast to glowing reports about Rechem in the financial press, people living near the incinerator paint a grim picture of its operations. Schoolteacher David Powell, a leading critic, called the plant "a fuming chemical cauldron in our valley." A report by the local waste disposal authority, the Borough of Torfaen, detailed an explosion, blowbacks, spills, and "a most disturbing increase" in PCB contamination near the plant. PCBs were found in grass and soil, as well as the Caldicotts' duck eggs.

Rechem specializes in destroying PCBs, which are highly injurious halogenated hydrocarbons linked to cancer, skin and gastrointestinal damage. These persistent chemicals can be reduced in volume by 99.999 percent when incinerated at high temperatures, but if the temperature gets too low, they give off poisonous dioxins and furans. Residents of Pontypool are angered that Rechem imports such dangerous substances into their town. An aggressive marketing campaign by Rechem brought PCBs from all over the world to the Pontypool facility.[8] Because many nations lack facilities for incinerating PCBs, Rechem's incineration business is expected to grow throughout the 1990s.

Because of the enormous global market for PCB disposal and the scarcity of facilities, Rechem can charge premium prices for its services. Imports make up an especially profitable part of the business: while imports made up only 13 percent by volume, they accounted for 35 percent of the revenue. As British pollution standards become more stringent and illegal dumping increasingly controlled, there will be only more work for companies like Rechem, predicted the financial newspaper, *Investors Chronicle.*

Rechem has aggressively pursued its interests and retaliated against some of its critics, including those who worry that Rechem is insufficiently regulated. In December 1989, a company executive revealed that the company, not the governmental agency responsible for enforcing environmental laws, did the monitoring for PCBs.[9] Rechem was not required to tell the Torfaen Borough Council the amount of pollution it was permitted to dump into public sewers; the agreement was confidential between the company and the Welsh water authority.[10] Further, Rechem took the Torfaen council to court for independently testing soil samples near its incinerator and announcing the findings. The company even obtained a temporary court injunction forbidding the council from taking further samples without company approval and banning further discussion of the council's findings.

Rechem has also filed about a dozen lawsuits against its critics, including a 1985 libel suit against the BBC for implying that its Pontypool plant was unsafe. The BBC apologized. Rechem also sued schoolteacher Powell for libel in 1989 over a radio interview in which he questioned the safety of the Pontypool incinerator.

Asked to comment on the controversy about the plant, Rechem spokesman Allan Woods said, "These are not simple matters; it would take hours to get it straight. You have given simple conclusions from one point of view." Woods ended with a warning to the authors: "You are walking down a dangerous road. We have taken action against media people who have grossly and maliciously libeled us."[11]

Having failed in their attempts to obtain stricter government regulation of the Rechem facility, Powell and other residents of Pontypool organized public protests outside the Belgian embassy in London in March 1990. The townspeople asked Belgium, a major supplier, to stop sending its toxic waste to Rechem. As part of the protest, Powell sent the Belgians excerpts from two journals kept by neighbors of the Rechem plant. An entry for January 13, 1990 noted: "White cloud over garden and both fields. Melting/burning polyethylene-type smell. Mrs. Caldicott had respiratory difficulty for one hour." On March 9, an entry noted "smoke emissions and smell beyond any sense."

London: Her Majesty's Impotent Inspectorate

If someone wrote a play about the plight of the environment in Great Britain, the playwright might be accused of melodrama if the main character in charge of enforcement, surrounded by critics and abandoned by his lieutenants, crept into his parked car, turned on the engine, and waited for the fumes to do their lethal work. But that is just what happened in early December 1989. Brian Ponsford, the chief of Her Majesty's Inspectorate of Pollution (HMIP), was reportedly so devastated by the failures of his agency that he committed suicide in his garage in Cricklewood, North London.

The beleaguered Ponsford could neither assemble the staff and resources needed to perform his job nor end the internal strife that plagued his agency. Formed in 1987, HMIP faced the impossible task of cleaning up the environment under a government that endorsed lax pollution standards and glorified the wonders of the free market. Several of Ponsford's chief inspectors had quit to work in private industry. Eight positions had been advertised but only one was filled. At the time, the U.S. EPA had 20,000 on staff to monitor air pollution and the Dutch inspectorate had 1,000, but Britain employed only 32 inspectors.[12]

Another event that occurred in 1989, very different from Ponsford's tragic death, though perhaps just as symbolic, could presage things to come in the United Kingdom. That year, trade union members, environmentalists, and angry residents of Liverpool and other cities forced Soviet tankers laden with PCBs from Canada to turn back rather than unload their toxic cargo in British ports. Greenpeace inflatable rafts attached a skull-and-crossbones poster to one Soviet freighter, and dock workers refused to handle the toxic wastes, which to Rechem's great annoyance were then hauled back across the Atlantic.[13]

One optimistic magazine writer, noting that the British Green Party attracted two million votes in 1989, heralded the demonstrations as marking "the year when the people of Britain said 'no' to being the dustbin of the world." That, however, seems at odds with the thinking of the nation's financial press. In an article headed "Where There's Muck, There's Brass," one paper called waste disposal a sure-fire growth industry for the 1990s.[14]

North of the Border

If the waste trade is flourishing in the United Kingdom, it is booming in Canada, and for this Canadians have Americans to thank. As the United States' foremost toxic waste trading partner, Canada takes in about 85 percent of all exported U.S. waste. Canada imports, processes, and exports hazardous wastes to both First World and Third World countries. Every day, truckloads of waste from such companies as Monsanto, Dow Corning, IBM, General Electric, and Ashland Oil cross into Canada, most of them bound for either Tricil Ltd., which operates a toxic waste incinerator in Sarnia, Ontario, or Stablex Canada Inc., a toxic waste processor with a large landfill site in Blainville, Quebec.

Tricil and Stablex earn tens of millions of dollars annually as America's dumping ground. The Ontario Ministry of Environment estimates that nearly 150,000 tons of toxic waste enter from the United States each year, while 46,000 tons leave Canada for disposal in the United States.[15] One component of the waste stream from the United States is spent batteries, which are shipped to Canadian smelters. As in Taiwan and elsewhere, lead from the smelting process is affecting workers and neighboring residents. A Mohawk high school next to a smelter in St. Catherine, Quebec, had to close down its garden because of high lead levels in the soil.

Canada's environmental standards are less strict than those in the United States, and the country offers relief from the "cradle to grave" responsibility for hazardous substances mandated by the Resource Conservation and Recovery Act. As Stablex advertises: "Once your industrial waste has been accepted by the Stablex facility for treatment, its safe disposal ceases to be your problem." The company implies that U.S. firms will be exempt from potential liability under Superfund legislation and thereby escape U.S. domestic disposal responsibilities. In its brochure, Stablex says it will "account for your waste by completing all required regulatory forms on your behalf."

Stablex, which took in more than 200,000 tons of hazardous waste from the United States from 1985 to 1989, mixes the toxic material with a cement-like substance, then buries the resulting slag in a landfill near Montreal. The Stablex landfill has no liners, as would be required in the United States, to prevent soil and groundwater contamination.[16]

The United States has a special bilateral agreement with

Canada, signed in 1986, requiring only that U.S. shippers declare what their waste is, where it is going and how frequent shipments will be. Although Canada can object, acceptance by that country is assumed. This fits nicely with current U.S. policy, which opposes any law prohibiting waste exports to a nation whose environmental standards are less stringent than those of the exporter. The Canadian government agrees, claiming that the United States ought not to infringe inadvertently on the sovereignty of another nation by extending American standards outside its own borders. When members of Congress introduced legislation to restrict the waste trade, Tricil and Stablex lobbied in Washington against the bill. The companies conceded that their operations did not meet U.S. specifications but argued against imposing "the American way of life" on Canada.

Schemes, Scams, and Smuggling

One U.S.-to-Canada waste enterprise was an illegal scheme that went undetected for four years. In Buffalo, New York, tanker trucks secretly filled the bottoms of their tanks with burnable chemical wastes—solvents and waste oil, some laced with PCBs. The drivers then filled the tanks' remaining capacity—about 90 percent—with diesel or heavy heating oil at refineries and delivered the tainted fuel at a bargain price to customers across the border in Ontario. Among the buyers were trucking companies, construction firms, asphalt plants, gas stations, and ultimately consumers.

Bootleg oil companies provided customers for the low-priced fuel. Canadian gas-station managers, using cellular phones, ordered the fuel from the bootleggers, who in turn called suppliers in Buffalo to arrange for shipments across the border. As another part of the scheme, bootleggers installed a long compartment in the top of tankers and filled it with oil dyed red to look like heating oil, which is not subject to tax. The concealed remainder of the load contained diesel oil, which was particularly profitable. "I make about $600 a load or about $15,000 a week," one bootlegger told a Canadian newspaper.

The scam succeeded because customs officials almost never inspected fuel trucks. Additionally, Canadian and U.S. monitoring systems for hazardous waste, which "work well if everyone involved in the disposal chain is honest," as a New York official noted, fail when they're not. Companies can easily falsify records. "You don't think we keep nice tidy records to help the tax man, do you?" laughed a Canadian bootlegger to a reporter.

Behind the truckers and their toxic loads was a sophisticated criminal enterprise that succeeded in smuggling several million gallons of toxic wastes into Canada over the four years. The bootleggers profited three ways: fees for disposing of the wastes (as much as $1,000 per drum of waste coolant or lubricant contaminated with PCBs), fees from selling the fuel, and evasion of the fuel tax owed to the Ontario provincial government, which amounted to about $100 million annually. Before the Toronto *Globe and Mail* broke the story of the operation in 1989, bootleggers had made millions of dollars and exposed unwitting Canadians to illegal levels of toxic contaminants.

After the story became public, Canadian officials admitted that they had known about the contaminated oil since 1987, yet had done nothing about the health hazards or law violations involved. Disposal experts said the tainted fuel would produce toxic emissions, including deadly furans and dioxins, when burned in truck engines or industrial boilers.

In the wake of heated criticism, Canadian government officials closed down 125 of 175 entry points for fuel tankers to facilitate future inspections. No further smuggling was detected, and in June 1990, the government's year-long investigation resulted in eighty-eight criminal charges of conspiracy, fraud, and theft against twelve men and women.[17]

"A Gigantic Poison Kitchen"

In the Germanys, the path of least resistance for the hazardous-waste stream led from the more prosperous, capitalist West to the poorer, socialist East. Hungry for foreign exchange, the East German government maintained a bank account in West Germany for depositing the funds earned by taking in waste. Data on environmental degradation were closely held state secrets, made public only after the collapse of the Communist government. Because political freedom was heavily circumscribed, no strong environmental movement had arisen to question the waste trade.

Statistics alone reveal what environmental carelessness has done to East Germany: 80 percent of its rivers are contaminated, most cities have air pollution fifty times the safe limits; six million people suffer from environmentally induced disease, and in some cities nine out of ten children suffer respiratory diseases.[18] Given such a consequence, another nation might choose to combat the worsening situation by disallowing further waste shipments—particularly after the repressive government whose policies had brought about the situation was removed from office.

But the relative poverty of East Germany—and its convenience to West Germany—may force it to continue to serve as a dumping ground. Despite the demise of the Communist government, environment official Marianne Montkowski said in November 1989 that the new government wanted to remain a major waste importer, explaining, "We need the foreign currency." She estimated that taking in foreign wastes would earn East Germany at least $50 million a year.

Four decades of cowboy dumping in East Germany have turned that country into what the magazine *Der Spiegel* termed "a gigantic poison kitchen." Just across the border from the West German town of Luebeck, the Schoenberg waste dump imports more than a million tons of waste a year, from household trash to dioxin-contaminated ash, at the low cost of $50 to $80 a ton. The ten-year-old dump is notorious for its lax environmental and safety standards: samples of wastes are examined only occasionally, and no barriers separate one kind of waste from another. A truck driver who used to dump material at Schoenberg told *Der Spiegel*, "The toxic stuff is leaking into the trash from private homes," adding, "I never saw a chemist or any person in charge of controls who might have cared about the content of the truck load I was delivering."[19]

Residents of Luebeck, about four miles away, fear their drinking water may be contaminated by leaching from the massive dump. Guenter Wosnitza, a local Green Party member who headed the Luebeck City Council committee on Schoenberg, suspects that "the worst kinds of wastes that could not be disposed of anyplace else in Europe end up in Schoenberg," and that many illegal shipments take place. Other sites of toxic dumping in East Germany include one specializing in contaminated metals at Ilsenburg, where soil is heavily polluted with dioxins and heavy metals; Schoineiche, near Berlin, where the base of the pit has not been sealed off and groundwater has been found to contain ammonium and sulfate; and the Vorketzin landfill near Potsdam, where contaminated soil was shipped.

The impending unification of the two Germanys has led to a complex situation. It has brought the environmentally devasted East home to the West. But it has also allowed East German environmentalists to organize. Said Katrin Zielke, a

Miles Decoster

Green Party member in East Germany, "We want to prevent the present reform movement in our society from building a self-advancing society of waste and disposable mentality under the pressure of unreasonable, short-sighted, materialistic need to catch up with the West."[20]

Armed with data about the waste traffic and environmental degradation released by the new regime, activists like Zielke stepped up public protests and won apparent victories. East Germany promised to ban the import of toxic waste from West Berlin, and Schoenberg promised to reduce the amount of toxic wastes it accepts. But as of June 1990, according to Greenpeace organizer Andreas Bernstorff of Hamburg, the promises had not been kept. Moreover, Bernstorff warned, West German industry and townships were rushing in with proposals to construct giant waste-incineration and chemical-processing plants in East Germany before stricter West German environmental laws are phased in fully by the year 2000. "Unification theoretically gives us more power to fight these projects," said Bernstorff, adding that if East German facilities don't follow West German law, environmentalists can file lawsuits to try to stop them. But the outcome, he predicts, will depend less on the courts than on the power and unity of the citizens and environmental movements in both Germanys.

The elected governments of the two Germanys have not followed the environmentalists' lead. While the Green Party called for a ban on exports to Third World countries, the governments have both called for the protection of the international waste trade. At the Basel Convention, East Germany aligned itself with West Germany against a coalition of Third World nations; the German diplomats fought for the right to make bilateral agreements outside the Basel treaty and opposed the demand of less-developed nations that exporters be held liable for damages their waste might cause to recipient nations.

The German situation, like that of Canada and the United Kingdom, illustrates the imperative of hazardous waste: it has to go somewhere, and the industrialized world seeks to keep many channels open. Now that the Berlin Wall has fallen, if West German and other European investors have their way, East Germany will be transformed from the simple dumping ground it was under Communism into a site for hazardous-waste incineration and processing. Some nations of the First World have indeed found in the waste trade a toxic gold mine.

**Discarded U.S. telephones—like these in Taiwan—
don't simply vanish with the garbage truck.**

SEVEN

Rethinking the Future

T hese days, garbage seems to be on everyone's mind. "Biodegradable" and "environmentally friendly" products of all types are appearing on supermarket shelves, and one major producer has even urged American consumers to purchase fewer goods as a way to cut down on waste. The news media have run story after story about the garbage issue, and in the late 1980s Hollywood took up the theme: a young homemaker in the film *sex, lies and videotape* had nightmares about America's trash problems; and in *Crimes and Misdemeanors*, Woody Allen fretted about the toxic-waste crisis.

Joel Hirschhorn was worrying about hazardous waste long before most of us realized that garbage doesn't simply vanish with the garbage truck.[1] In 1981, while working as a senior associate with the Office of Technology Assessment of the U.S. Congress (OTA), Hirschhorn tried to determine how much hazardous waste was being produced by U.S. industry. At that time, the EPA's official figures held that American industry was generating approximately 40 million tons a year.

A two-year investigation by Hirschhorn, however, determined that the total was at least 250 million tons a year— over six times the EPA figure—and in response to his report, the regulatory agency upgraded its number to the higher figure. In 1989, Hirschhorn said, the estimate doubled again—to about 500 million tons a year—and no one puts too much stock in that figure, either.

"We don't do a very good job measuring and knowing accurately how much waste is being generated in the first place," said Hirschhorn, who began investigating the hazardous-waste issue ten years ago, and is now one of the country's foremost waste experts. Today, he is president of EnviroSearch-East, an independent environmental consulting firm based in Washington, D.C.

Noting that a great deal of industrial waste that is in fact hazardous is not currently regulated as such, Hirschhorn suspects that the total amount of hazardous waste generated each year in the United States may in fact measure in the *billions* of tons. Given the fogginess of such basic statistics, it is not surprising that experts are unclear as to exactly how much of our hazardous waste is being dumped overseas.

The astonishing ignorance of the amounts of waste involved, along with regulatory loopholes and lax enforcement, have combined to foster a friendly business climate for operators like the Colbert brothers, the owners of the "poison ships," and other toxic traders. Government officials concede that the problem of waste exports continues to be a serious one: "Waste export is not only an environmental issue, it's a foreign policy issue," said Wendy Grieder, a specialist with the International Activities Office of the EPA. ". . . We've come very close to having major difficulties with countries because of our exports."

But over the long term, tighter controls on exports will not be enough: government action alone cannot solve the problem of the global dumping ground, which feeds on the dizzying amount of waste produced worldwide each year. The solution is at once simple and overwhelmingly difficult: industrial societies must stop producing so much waste.

As long as waste can be sent overseas, corporations can

avoid the tougher measures necessary to cut down on the use of harmful chemicals and to reduce waste production at home. "[Exporting] only encourages industry to churn out more and more waste every day rather than preventing it at the source," said Jim Vallette, an activist with the environmental organization Greenpeace.[2]

Greenpeace first began looking into the international toxic trade in 1987, Vallette explained, when it was discovered that waste producers, under pressure to stop dumping toxic chemicals near neighborhoods and wilderness areas, were beginning to ship them overseas. So that Greenpeace's "victories in the North wouldn't translate into defeats in the South," as Vallette put it, the organization launched a campaign to stop waste exports.

"As long as this escape valve exists," he warned, "waste reduction is going to be pushed further and further into the future."

Not surprisingly, companies that generate hazardous waste want to hold on to the waste-export "safety valve." In the United States, such firms have opposed the Waste Export Control Act, which would (among other things) permit foreign governments to sue U.S. exporters for damage caused by waste shipments.[3] In its opposing testimony, the U.S. Chemical Manufacturer's Association (CMA) stated, "In the CMA's view, the existing system of export controls for hazardous wastes works."[4] At the same time, the Scrap Metal Recycling Institute opposes controls on exports because it believes scrap dealers should be able to sell where prices are best.

But Vallette, Hirschhorn, and an increasing number of experts say there is a better solution to the waste crisis. Reducing waste at its source, they argue, is by far the most effective way to solve our waste problems and limit environmental pollution, as well as conserve energy and vital resources. Unfortunately, source reduction has been neglected in favor of other waste-management schemes. As Louis Blumberg and Robert Gottlieb have pointed out in their thought-provoking book *War on Waste*,[5] the free market dictates that waste-reduction decisions be left up to the private sector. And all too often, waste production is good business in the short run.

Blumberg and Gottlieb have documented the steady increase in global waste production during the last half century and emphasized how large a part of our economy it has become. Hazardous waste is generated in the production, packaging and disposal of some of our most basic consumer items. Even innocuous household garbage, when incinerated, may release dioxins and other compounds, and end up as toxic ash. As much as one dollar of every ten that Americans spend for food and beverages goes for packaging alone.[6] This packaging material, made increasingly of plastic, accounts for as much as one-third of all municipal waste in the country. The average consumer now goes through sixty pounds of plastic packaging a year, fueling an industry that uses some of the most highly toxic chemicals in the world and generates waste at every step of the production process.

In addition, the shift from paper to plastic since World War II leaves us with compounds that cannot be recycled and whose half-lives in the environment can be measured in the hundreds of years. The effects go beyond simple poisoning: an estimated 100,000 marine mammals die needlessly each year from plastic packaging debris such as the six-pack rings discarded by consumers of soft drinks and beer.

After source reduction, reuse is the next most efficient way to minimize waste production. It's an old-fashioned idea: forty-five years ago it was common practice for beer and soft-drink bottles to be reused as many as thirty or forty times. But reuse and source-reduction efforts—that is, waste *prevention*—have received little support from either government or industry, which have focused instead on ways to dispose of waste after it is generated: constructing incinerators, finding more space for landfills, and permitting waste to be sent overseas.

At the same time, reduction and reuse have received less attention than recycling, which is a far less efficient option. A bottle that is reused, for example, is simply cleaned and refilled, while a recycled bottle must be crushed and refashioned, requiring more energy and resources. And when it comes to batteries, computers, and other products containing hazardous materials, recycling clearly poses environmental and health hazards if it is done under unsafe conditions. According to Joel Hirschhorn: "[recycling] is not the same as not producing the waste to begin with. Recycling is a form of waste management. In my hierarchy of values, I see at the very top true prevention."

Hirschhorn is convinced that efficient use of resources is not incompatible with a thriving economy—on the contrary. "The greatest misconception I find among all sorts of people is the assumption that we have to produce toxic waste," he said. "And that is simply wrong. We can run American industry, we can operate industrial facilities, we can make the kinds of products people want, and we don't have to produce so much toxic waste. . . . If you look at successful countries like Switzerland and Germany and Japan, one of the reasons [their industry] is more competitive and more profitable than American industry is the fact that they're more efficient at what they do and produce less waste. Waste is inefficiency. Waste is non-competitive."

Considerable blame for our hazardous-waste problems has been placed on consumers, particularly in the United States, because of their supposed infatuation with time-saving and disposable products—no matter what the cost. While our grandparents saved twine and jars to use over again, and carried their groceries home in a basket, modern-day consumers are accustomed to buying products swathed in plastic, aluminum foil, and Styrofoam, to using a razor or a battery or even a camera just a few times and then tossing it into the trash. But it is unclear whether people in modern industrial societies have had much choice in the evolution of their consumer culture, with its abundance of products and its emphasis on planned obsolescence. Consumer demand for convenience has in fact been shaped by producers through a ceaseless barrage of advertising and marketing pressure. Given a choice, Americans might not choose to pay for a throwaway society with their health or the integrity of the global environment.

"Consumer habits are actually very flexible," said Peter Gleick, a director of the Pacific Institute for Studies in Development, Environment, and Security. "Consumers aren't demanding Styrofoam or CFCs [chlorofluorocarbons], for example. They just want something to put their coffee in."[7]

In fact, citizens have begun to protest the blizzard of Styrofoam and plastic that accompanies each trip to the store or fast-food restaurant. After a number of lean years, diaper services are seeing business grow again as many parents, citing environmental concerns, choose cloth over disposable diapers. Computer companies are also noticing changes: an IBM official said customers are increasingly concerned that the company dispose of used computers in an environmentally sound way.

At the same time, McDonald's, America's largest user of the foamed polystyrene "clamshell" containers (the company generates 70 million pounds of discarded materials a year) has become the object of several consumer boycotts in recent years. One group, "Kids Against Pollution," staged a protest at the United Nations in New York, featuring students dressed up as "Ronald McToxic." The young protesters admonished passersby to stop eating at the restaurant, chanting, "The planet deserves a break today!"

The kids' crusade against McDonald's is "part of an environmental tidal swell," argued James Ridgeway and Dan Bischoff in the weekly *Village Voice.* The change in attitudes, they wrote, "is forcing a major reassessment of priorities from the boardrooms of America's premier chemical concern, DuPont, to the roadside takeout windows of the suburbs. And this movement is not being led by Washington-based environmentalists—who would never dream of eating at McDonald's—but by people who *do* eat there all the time."[8]

Some U.S. companies have taken important steps to cut back on the amount of hazardous waste they produce. The 3M company, for example, which makes floppy discs and other products, as well as adhesive tape, has reported savings of about $300 million over a decade by implementing preventive techniques that reduced waste generation by 50 percent. The savings have resulted in a 5 percent boost in the company's net income.

Efforts like these are applauded by researchers at INFORM Inc. a New York-based, non-profit research organization seeking innovative ways to solve environmental problems. In 1985, INFORM studied twenty-nine chemical plants and found that remarkable possibilities exist for reducing much of the hazardous waste they generate, and that it usually makes good business sense to do so.[9] Although INFORM's study covered U.S. companies only, its findings seem clearly applicable overseas.

Borden Chemical Company reported that its plant in Fremont, California, which manufacturers industrial resins and adhesives, was able to reduce organic contaminants in its waste water by 93 percent through a series of simple changes. Before these steps were taken, says Frank Tejera, Borden's plant manager, the company's waste was going into an evaporation pond at the facility. Tejera saw the way the regulatory winds were blowing and knew the company would have to do something about the waste—which would mean dealing with skyrocketing disposal costs. Since the company cut its waste, the pond has been cleaned; today ducks are living there. "The pond used to be pretty ugly, full of nasty-looking stuff," says Tejera. "It was attracting attention—the wrong kind. Even the ducks knew better than to get near it."

Another company, USS Chemicals, produces a variety of chemicals at its Haverhill, Ohio, plant—a formidable facility of tanks and spindly towers not far from the Ohio River. The company was able to save an estimated $100,000 in raw material costs by reducing air emissions by 400,000 pounds per year.

INFORM's report also contained some bad news, however. The success stories it discussed represent only a "minute fraction of the billions of pounds of wastes" that the plants studied generate each year. The organization identified the patchwork nature of government regulatory schemes as a factor that discourages waste reduction by companies. Unfortunately, it is often cheaper in the short run for firms to comply with disposal rules than to practice the kind of overall efficiency that would benefit everybody, including the producer, in the longer term. According to the report: "Waste reduction alternatives were seldom considered until circumstances virtually forced plants to review their waste management practices. Significantly . . . once waste reduction measures were adopted, they were found to save the plants money."

In its 1986 report, "Serious Reduction of Hazardous Waste,"[10] the OTA also found substantial savings by companies that adopted waste-reduction measures. But the study, directed by Joel Hirschhorn, pointed out that "although there are many environmental and economic benefits to waste reduction, over 99 percent of Federal and State environmental spending is devoted to controlling pollution after waste is generated. Less than 1 percent is spent to reduce the generation of waste."

As John O'Connor of the Boston-based National Toxics Campaign has observed, our environmental laws "never tell industry that it mustn't make products or use processes that create pollution. Rather, these laws say that once pollution is created, firms must put filters on the ends of their pipes to capture the sludge, dust, gas, or fumes. The filters take poison out of the air only to be dumped on the land in the form of toxic sludge. Workers are protected simply by moving the poison fumes from the shop floor out into the community through ventilation." Concluded O'Connor, "The guiding logic seems to be, take it out of the air and put it in the water, but never challenge the making of poisons in the first place."[11]

This flaw in our overall approach to the waste crisis troubles experts like Hirschhorn. "We have many people in the United States government [with] a wrong-headed position, who are mostly concerned about finding a place to put toxic waste. Now if we can find a country to take our toxic waste, there are some people in the government who think that means [it] helps American industry." Instead, said Hirschhorn, "what helps American industry is technical assistance and research support to produce less waste in the first place."

And, as Hirschhorn pointed out, what applies to the United States is also true worldwide. The stakes are considerable: "If the rest of the world aspires to our way of life . . . we'll bury ourselves in waste," he said. "There literally is no escape from pollution, and a global society that tolerates pollution is a global society that is threatening its own survival."

Like radioactive waste, the hazardous waste already generated by modern society will be with us for generations, and the best technology available must be used to treat it as safely as possible—both in the developed and developing worlds. But Hirschhorn and other experts say there are virtually no incinerators, no landfills, and no known waste-disposal methods that do not release pollutants. As one analyst put it, the term *waste management* "is an oxymoron."

And for now, at least, "waste management" still includes shipping hazardous waste overseas, along the path of least resistance—typically from wealthy to poorer regions. Accordingly, environmental groups and much of the Third World have called on all countries to ban the export of hazardous wastes. Why?

"On the grounds that you don't poison your neighbor," explains Jim Vallette of Greenpeace. "You don't dump your garbage on your neighbor's lawn for moral reasons. It's that simple."

Videographer: Carl Gilman

**Mexican children bathing in
contaminated water.**

Notes

Chapter 1

1. *U.S. News and World Report*, 21 November 1988; *Newsweek*, 8 November 1988; *New Straits Times*, 26 July 1988; and other sources.
2. Muhammadu Jibirilla, Information Attache, Consulate of Nigeria, Washington, D.C., interview by Stephen Levine, 1 August 1990.
3. U. S. Environmental Protection Agency, Report of Audit E1D37-05-0456-80855, "EPA's Program to Control Exports of Hazardous Waste," 31 March 1988.
4. Frank D'Itri, Department of Fisheries and Wildlife, Michigan State University, interview by Stephen Levine, July 1990; *St. Louis Post-Dispatch*, 26 November 1989.
5. Harry Perks, Streets Commissioner of Philadelphia, quoted in *Philadelphia Inquirer*, 9 June 1986.
6. "The Global Poison Trade," *Newsweek*, 7 November 1988, 66.
7. Unreleased draft of *National Survey of Hazardous Waste Treatment Storage, Disposal and Recycling Facilities: Final Report*, submitted to the Environmental Protection Agency by Research Triangle Institute (Durham, NC); Kate Blow, U.S. Environmental Protection Agency, interview by Leslie Haggin, 30 July 1990. The Resource Conservation and Recovery Act (RCRA) defines as "hazardous" materials that may "pose a substantial

115

present or potential hazard to human health or the environment when improperly treated, stored, transported, disposed of, or otherwise managed."

8. The Global Tomorrow Coalition, Walter H. Corson, ed., *The Global Ecology Handbook: What You Can Do About the Environmental Crisis* (Boston: Beacon Press, 1990), 246.

9. California Waste Management Board, Technical Information Series, chap. 3a, "Waste to Energy."

10. U. S. Environmental Protection Agency, Office of Solid Waste and Emergency Response, *The Waste System*, (Washington, DC: U.S. Government Printing Office, November 1988), 1–20.

11. *The Economist,* 8 April 1989, 24.

12. June Juffer, "Dump at the Border," *The Progressive,* October 1988, 24.

13. Rosa Devora, resident of Ejido Chilpancingo, Mexico, interview by Bill Moyers, 28 April 1990.

14. Andrew Porterfield and David Weir, "The Export of U.S. Toxic Wastes," *The Nation,* 3 October 1987.

15. H. Jeffrey Leonard, "Hazardous Wastes: The Crisis Spreads," *National Development,* April 1988, 43.

16. Hilary F. French, "A Most Deadly Trade," *World Watch,* July–August 1990.

17. *San Jose Mercury News,* 25 August 1988.

18. *Africa Report,* September–October 1988, 48.

19. Andrew Porterfield, "To Tonga With Love," *California Business,* December 1987, 68.

20. *Africa Report, op. cit.,* 48.

21. Bill Moyers, *A World of Ideas* (New York: Doubleday, 1989), 333.

Chapter 2

1. Mark Jaffe, "Ash is gone, but freighter continues its troubled course." *The Philadelphia Inquirer,* 28 October

1989, 1-B. Jaffe's reporting on the *Khian Sea* from 1986–1989 was a key resource for this chapter.

2. Rep. Michael Synar (D-Okla), quoted in Mark Jaffe, "Tracking the Khian Sea," *Philadelphia Inquirer*, 15 July 1988, 1-B.

3. Jim Vallette, Greenpeace, quoted in "The Global Poison Trade," *Newsweek*, 7 November 1988, 66.

4. John C. Martin, "Philadelphia Incinerator Ash Exports for Panamanian Road Project—Potential Environmental Damage," U.S. Environmental Protection Agency Flash Report, 5 October 1987. Subsequent references to Martin are also from this report.

5. Congress, House, Committee on Government Relations, *International Export of U.S. Waste: Hearing before a Subcommittee of the Committee on Government Operations*, 100th Cong., 2nd Sess., 14 July 1988.

6. The Bureau of National Affairs, "Environment Reporter: Current Developments," Database: BNA-ENV, 28 August 1987. Reports from this database, August 1987 through November 1988, were a key resource for this chapter.

7. Mark Jaffe, "Haiti is the latest to reject city ash," *The Philadelphia Inquirer*, 2 February 1988, 4-B.

8. "Evening News," CBS Affiliate WCAU-TV, Philadelphia, 2 February 1988.

9. Henry Dowd, former vice-president of Amalgamated Shipping, interview by William Kistner, Orlando, Florida, 2 April 1990. Subsequent quotes by Henry Dowd are also from this interview.

10. Mark Jaffe, "Philadelphia ash: Island leaders wonder what's next," *Philadelphia Inquirer*, 22 February, 1988, B-1.

11. Carol Morello, "Panama rejects city ash," *Philadelphia Inquirer*, 11 September 1987, B-1.

12. Ramona Smith, " 'Rust Bucket' Still Being Held at Bay," *Philadelphia Daily News*, 1 March 1988, A-3.

13. Edward Roe, port captain, U.S. Coast Guard, telephone interview by William Kistner, 19 December 1989; Ramona Smith, "Ship Tales: Explaining the One that Got Away," *Philadelphia Daily News*, 15 July 1988, A-20.

14. "The Global Poison Trade," *Newsweek*, 7 November 1988, 68.

15. In correspondence from Francesco Rizzuto, Italian lawyer representing the captain of the *Zanoobia*, to CIR reporter William Kistner, 11 April 1990: "I suggest also to ask the Drug Enforcement Agency in Washington to investigate in connection with Mercantil Lemport S.A. in Panama and a Mr. Luciano Miccioche, born at Ravanusa, Sicily, now a resident in Medellín, Colombia. . . . Mr. Micchioche was the agent of Jelly Wax in Venezuela. He escaped first to Panama and then to Medellín, Colombia, when the scandal blew out. They are involved also in the traffic of cocaine from South America to Italy and, as a matter of fact, a Mr. Vittorio Arullani, legal advisor of Jelly Wax, has been put in jail in Italy for illegal traffic in narcotics."

16. Ava Zunino, "Ma i fusti sono ancora tutti lì [But the drums are all still there]," *Il Lavoro*, 28 April 1990.

17. Carol Cirulli, "Toxic Boomerang," *The Amicus Journal*, Winter 1989, 11.

Chapter 3

1. Bruce Comfort, New Jersey Department of Environmental Protection, telephone interview by Andrew Porterfield, 1987.

2. David Wilma, U.S. Environmental Protection Agency, telephone interview by Loren Stein, 20 July 1990.

3. *Resource Conservation and Recovery Act of 1976* ("RCRA"), P.L. 94-580, 42 U.S. Code 6901 et seq.

4. Jim Vincent, U.S. Environmental Protection Agency, telephone interview by Loren Stein, 18 July 1990.

5. Freedom of Information Act (FOIA) documents obtained from the U.S. Environmental Protection Agency, 1987; anonymous source, North Carolina Department of Agriculture, telephone interview by Andrew Porterfield, 1986.

6. Indictment, 1986, U.S District Court, District of New Jersey.

7. The U.S. Senate passed legislation in 1990 as part of its farm bill that would ban the export of agricultural chemicals illegal in the United States.

8. Agency for International Development memorandum, 1987.

9. James R. DeVita, interview by Bill Moyers, New York, 8 March 1990; Allen P. Rossi, interview by Bill Moyers, New York, 8 March 1990.

10. Allen P. Rossi, interview by Bill Moyers, New York, 8 March 1990.

11. Government Sentencing Memorandum, 85 Cr. 1134 (CLB), 1986, 10.

12. *Mount Vernon Daily Argus*, Westchester Rockland Newspapers, White Plains, New York, 4 June 1983.

13. Office of the County Prosecutor, Essex County, videotape, Newark, N.J., 1986.

14. Government Sentencing Memorandum, 85 Cr. 1134 (CLB), 1986, 6.

15. Government Sentencing Memorandum, 85 Cr. 1134 (CLB), 1986, 12–13; Indictment, 1986, United States District Court, District of New Jersey; Indictment, 14 August 1986, State of New Jersey v. Colbert Brothers, et al.; Indictment, 1984, Superior Court of New Jersey, Essex County (Criminal), The State of New Jersey v. Signo Trading International, Ltd., et al.

16. Various state and federal environmental officials, interviewed by Loren Stein, July 1990.
17. Paul Jourdan, Institute of Mining Research, University of Zimbabwe, Harare, interview by Loren Stein, 31 July 1990.

Chapter 4

1. Roy Akridge and Jim Purser, U.S. Customs Service, interview by William Kistner, Otay Mesa, 6 February 1990. Other U.S. Customs Service citations are also from this interview.
2. Sgt. Lance Erickson, California Highway Patrol, interview by William Kistner, Los Angeles, 6 February, 1990; interview by Sarah Henry, Los Angeles, 23 July, 1990.
3. Andrew Porterfield, "To Tonga, With Love," *California Business*, December 1987, 68–71.
4. William Carter, deputy district attorney, environmental crimes division, Los Angeles County District Attorney's Office, interview by Lowell Bergman, Los Angeles, 29 May 1990; interview by Sarah Henry, Los Angeles, 24 July 1990. Subsequent quotes from Carter are also from these interviews.
5. Jorge Eulloqui, Barmet Aluminum Torrance division plant manager, interview by Lowell Bergman, Torrance, 17 May 1990.
6. David Torres, interview by Lowell Bergman, Burbank, 23 March 1990.
7. The People of the State of California v. Melvyn Ingalls, Albert Mangrum and Does 1 through 50 inclusive, Superior Court of California, County of San Diego, Complaint for Injuction, Civil Penalties and other Equitable Relief, Case No.: 488199, 23 June 1982; Final Judgment, Case No.: 488199, 16 June 1986. Also, United States of

America v. Melvyn Ingalls, Albert Mangrum, Herman Avilez, Jack E. Rust and Industrial International Corp., United States District Court, Southern District of California, Consent Decree, Civil Action No.: 85-0282-GT (IEG), 17 October 1986.

8. Jack Rust, senior environmental consultant, Valley Environmental Services, interviews by Lowell Bergman, San Diego, 22–23 March 1990; interview by Bill Moyers, San Diego, 28 April 1990; interview by Sarah Henry, Burbank, 24 July 1990.

9. United States of America v. Raymond Franco and David Torres, United States District Court, Central District of California, Indictment, November 1989, Grand Jury Case No.: CR 90-352. Subsequent references are also to the Franco-Torres indictment.

10. Robert Brosio, Acting United States Attorney, Los Angeles, press release, 10 May 1990.

11. Data obtained from Lawrence Jackson, senior waste management engineer, chief of the Land Disposal Restrictions Unit, California Department of Health Services. Source: Department of Health Services Hazardous Waste Information System (HWIS) database.

12. Marc Lifsher, "Neighbor frets about newspaper-waste dump." *The Orange County Register*, 13 February 1986, A1, A14.

13. David Takashima, aide to California legislator Steve Peace, telephone interview by Sarah Henry, 11 July 1990.

14. Jane Juffer, "Dump at the Border: U.S. Firms Make a Mexican Wasteland," *The Progressive*, October 1988, 24–9.

15. The agreement referred to is the Agreement Between the United States of America and the United Mexican States on Cooperation for the Protection and Improvement of the Environment in the Border Area (La Paz, Baja California, Mexico, 14 August 1983). The later annex

referred to is Annex III, the Agreement of Cooperation
Between the United States of America and the United
Mexican States Regarding the Transboundary Ship-
ments of Hazardous Wastes and Hazardous Substances
(Washington, D.C., 12 November 1986). Notification
statistic obtained from Jeff Inglis, Office of Health and
Emergency Planning, U.S. Environmental Protection
Agency, interview by Elana Rosen, 30 July 1990.
16. Patrick McDonnell, "Border Boom Feeds Hazardous-
Waste Ills," *Los Angeles Times*, 10 September 1987 (San
Diego edition), II-1.
17. Lowell Bergman, "The Dirtiest River," CBS "60 Minutes" (transcript), 28 December 1986.
18. Maria de Consuelo Montoya, Rosa Devora, and Dr. Juan
Manuel Sánchez León, interviews by Bill Moyers, Ejido
Chilpancingo, Mexico, 28 April 1990.
19. Soil and water sampling and analyses, tests conducted
by Jim Smith, environmental scientist, Ejido Chilpancingo, Baja Mexico, 28 April 1990.
20. Jorge Escobar Martínez, chief, Department of Prevention and Control of Environmental Contamination, Secretaría de Desarrollo Urbano y Ecología (SEDUE),
Mexicali, interview by William Kistner, 10 November
1989.

Chapter 5

1. Former Tonolli workers Manuel Pereira de Deus, Jose
Balbino de Oliveira, and others, interviews by Mike
Kepp, January and July 1990.
2. Bureau of the Census, U.S. Department of Commerce,
"U.S. Exports of Domestic and Foreign Mechandise by
All Methods of Transportation (Lead Waste and

Scrap)," schedule B, in *Statistical Classification of Domestic and Foreign Commodities*, (Washington, D.C.: U.S. Government Printing Office, 1989); *Recycling Lead: Responding to the Challenges of the Late 1980s*, 146–7 (London and New York: Commodities Research Unit Limited, 1987).

3. Wendy Grieder, interview by Bill Moyers, Washington, D.C., 7 June 1990.

4. James G. Palmer and Michael L. Sappington, *A Cleaner Environment: Removing the Barriers to Lead-Acid Battery Recycling*, Appendices B and C (St. Paul: October 1988).

5. *Philadelphia Inquirer*, 22 October 1987; Donna McCartney, Environmental Protection Agency, interview by Dan Noyes, 1 August 1990.

6. U.S. Government Accounting Office, *Superfund: A More Vigorous and Better Managed Enforcement Program is Needed*, GAO/RCED-90-22 (Washington, D.C.: U.S. Government Printing Office, December 1989).

7. HESIS Fact Sheet No. 4, Health Evaluation System & Information Service (Berkeley, October 1989), 1-2..

8. Christopher Norwood, *At Highest Risk* (New York: Penguin Books, 1981), 106–7.

9. *San Francisco Examiner*, 11 January 1990, A-11.

10. Suds-R.58, São Jose dos Campos, Ambulatorio de Doenças Provocadas Pelo Trabalho, Intoxicações por Chumbo [Emergency Room for Work-Related Illnesses, Lead Poisonings], 26 October 1989.

11. Mike Kepp, interview by Dan Noyes, 28 July 1990.

12. *Vale Paraibano*, Brazil, 5 November 1988, 1.

13. Mike Kepp, interview by Dan Noyes, 28 July 1990.

14. Dames & Moore, report prepared for the Center for Investigative Reporting, 29 June 1990.

15. James G. Palmer and Michael L. Sappington, *op. cit.*, p. 5 and Appendix A.

16. Steering Committee, Taiwan 2000 Study, *Taiwan 2000:*

Balancing Economic Growth and Environmental Protection (Taipei, 1989), 11.

17. Wayne Lo, Thai Ping Metal Industrial Company, interview by Dan Noyes, 12 April 1990; *American Metal Market*, 22 March 1990.

18. Eugene Chien, Environmental Protection Administration for the Republic of China, interview by Lowell Bergman, 9 April 1990.

19. Joseph Chen, interview by Lowell Bergman, 14 April 1990.

20. *Ibid*.

21. U.S Environmental Protection Agency memorandum, San Francisco, 16 May 1988.

22. Bureau of the Census, U.S. Department of Commerce, *op. cit.*

23. Global Telesis Corporation notification to the United States Environmental Protection Agency, Walnut Creek, California, 22 December 1988. Document obtained under the Freedom of Information Act (FOIA).

Chapter 6

1. Torfaen Borough Council, Presentation to the Secretary of State for Wales on the Operation at Rechem International Ltd. and the Need for Public Inquiry, 1989, Appendix 2, p. C.

2. Ken Caldicott, interview by Eve Pell, 28 July 1990; Shirley and Ken Caldicott, interview by Stephen Levine, August 2, 1990.

3. "An Atmosphere Poisoned by Mistrust," *The Independent*, 3 October 1989.

4. Geoffrey Lean and Eileen MacDonald, "Britain's Dirty Business," *Observer*, 4 September 1988, 15.

5. Tyler Marshall, "West Europe Has Its Fill of Toxic Waste," *Los Angeles Times*, 28 February 1989, 1.
6. Greenpeace policy statement, The International Waste Trade in the United Kingdom.
7. Lean and MacDonald, *op. cit.*, 15.
8. House of Commons Paper 80, Welsh Affairs Committee, Second Report: Rechem Limited Incineration Plant, Pontypool, 6 June 1990, xiv.
9. Paul Brown, "Rechem Denies Test Manipulation," *The Guardian*, 14 December 1989.
10. Borough of Torfaen, *op. cit.*, 9.
11. Allan Woods, interview by Eve Pell, 2 August 1990.
12. Andrew Morgan and Christopher Elliott, "Pollution Agency in Chaos as Staff Quits," *The Sunday Correspondent*, 10 December 1989.
13. Peter Beaumont et al., "The Week the Greens Went Overboard," *Observer*, 13 August 1990.
14. "Where There's Muck, There's Brass," *Investors Chronicle*, 28 October 1988.
15. Andrew Porterfield and Jock Ferguson, "U. S. Bill Could End Waste Flow to Canada," Toronto *Globe and Mail*, 18 August 1989, A8. Canadian environmental agencies' estimates of U.S. toxic-waste exports to Canada are higher than those of U.S. environmental agencies.
16. *Environmental Law Reporter*, 20 ELR 10062; Pierre Grenier, president of Stablex, interview by Eve Pell, 2 August 1990.
17. Jock Ferguson with Andrew Porterfield, Toronto *Globe and Mail*, series of reports on toxic fuel exports from the U.S. to Canada, 8–10 May 1989.
18. "Eastern Europe's Big Cleanup," *Business Week*, 19 March 1990, 114; Larry Tye, "The Scars of Pollution: Iron Curtain Rises to Reveal Dirt, Death," *Boston Globe*, 17 December 1989.

19. *Der Spiegel*, 18 July 1983.
20. Katrin Zielke, "The Green Curtain," *Mother Jones*, April/ May 1990, 53.

Chapter 7

1. Joel Hirschhorn, interview by Bill Moyers, Atlantic City, 6 June 1990.
2. Jim Vallette, interview by Bill Moyers, Atlantic City, 6 June 1990.
3. U.S. Representatives Synar, Conyers, Porter, and Wolpe, Waste Export Control Act, pending legislation H.R. 2525, Waste Export Control Act, introduced 31 May 1989.
4. Written Statement of the Chemical Manufacturers Association on the Waste Export Control Act, before the Subcommittee on Transportation and Hazardous Materials, Committee on Energy and Commerce, U.S. House of Representatives, 27 July 1989, 3.
5. Louis Blumberg and Robert Gottlieb, *War on Waste: Can America Win Its Battle with Garbage?* (Washington, D.C.: Island Press, 1989).
6. Cynthia Pollock, *World Watch Paper 76, Mining Urban Wastes: The Potential for Recycling* (Washington D.C.: The World Watch Institute, 1987), 8.
7. Peter Gleick, interview by Constance Matthiessen, Sharon Tiller, and David Weir, Berkeley, 21 August, 1989.
8. James Ridgeway and Dan Bischoff, "Fighting Ronald McToxic," *Village Voice*, 12 June 1990, 30.
9. David Sarokin, Warren Muir, Catherine Miller, and Sebastian Sperber, *Cutting Chemical Wastes: What 29 Organic Chemical Plants Are Doing to Reduce Hazardous Waste* (New York: INFORM, 1985).

10. U.S. Congressional Office of Technology Assessment, "Serious Reduction of Hazardous Waste" (Washington, D.C.: U.S. Government Printing Office, September 1986), 3.

11. John O'Connor, "Toxic Logic: Government 'Manages' Pollution into the Environment, But Never Stops It at Its Source," *Mother Jones*, April/May 1990, 49.

Resource Guide

Books, Documents, and Articles

Hazardous Waste: International Traffic

Batstone, Roger, James E. Smith, Jr., and David Wilson, eds. *The Safe Disposal of Hazardous Wastes: The Special Needs and Problems of Developing Countries. A joint study sponsored by the World Bank, the World Health Organization (WHO) and the United Nations Environment Programme (UNEP).* World Bank Technical Paper 0253-7494, no. 93; 3 volumes. Washington, D.C.: World Bank, 1989.

Broder, John M. "U.S. Military Leaves Toxic Trail Overseas." *Los Angeles Times*, 18 June 1990, A-1.

Brown, Lester, ed. *State of the World 1989: A World Watch Institute Report on Progress Toward a Sustainable Society.* New York: Norton, 1989.

Castleman, Barry I. "The Export of Hazardous Factories to Developing Nations." *International Journal of Health Sciences* 9 (4): 569–606 (1979).

Ceppi, Jean-Philippe, ed. *Nos Déchets Toxiques. L'Afrique a faim: "V'là nos poubelles!"* [Our Toxic Wastes. Africa is hungry: "Here is our garbage!"]. Le Bureau de Reportage et de Recherche d'Informations (BRRI). Lausanne, Switzerland: Edition du CETIM (Centre Europe-Tiers Monde), 1989.

Christrup, Judy. "Clamping Down on the International

Waste Trade." *Greenpeace* 13 (November/December 1988): 8–11.

Cirulli, Carol. "Toxic Boomerang." *The Amicus Journal* (Winter 1989): 9–11.

Corson, Walter H., ed. *The Global Ecology Handbook: What You Can Do About the Environmental Crisis.* The Global Tomorrow Coalition. Boston: Beacon Press, 1990.

Costner, Pat, ed. *Waste Traders Target the Marshall Islands.* Washington, D.C.: Greenpeace Pacific Campaign, 1989.

European Parliament. Secretariat. *Community Policy Concerning the Management of Dangerous Waste.* Research and Documentation Papers 11/06-1987. Luxembourg: European Parliament, Directorate General for Studies, 1987.

Forester, William S. and John H. Skinner, eds. *International Perspectives on Hazardous Waste Management: A Report from the International Solid Wastes and Public Cleansing Association (ISWA) Working Group on Hazardous Wastes.* London: Academic Press, 1987.

French, Hilary F. "A Most Deadly Trade." *World Watch* 3 (4): 11–7 (July/August 1990).

French, Hilary F. "Combatting Toxic Terrorism." *World Watch* 1 (5): 6–7 (September/October 1988).

Galli, Craig D. "Hazardous Exports to the Third World: The Need to Abolish the Double Standard." *Columbia Journal of Environmental Law* 12 (1): 71–90 (1987).

Handley, F. James. "Export of Waste from the United States to Canada: The How and Why." *Environmental Law Reporter* 20 (2): 10061–6 (February 1990).

Handley, F. James. "Hazardous Waste Exports: A Leak in the System of International Legal Controls." *Environmental Law Reporter* 19 (4): 10171–82 (April 1989).

Harland, David. *The Legal Aspects of the Export of Hazardous Products.* Penang, Malaysia: International Organization of Consumers Unions, 1985.

Harris, Tom and Jim Morris. "Uncle Sam's Hidden Poisons."

Sacramento Bee. Reprinted from a series published 30 September to 5 October 1984.

Hilz, Christoph. "Toxic Waste Exports to the Third World: An Analysis of Policy Options and Recommendations for an International Protocol to Protect the Global Environment." Ph.D. diss. Paper No. HSMP 16. Massachusetts Institute of Technology, Center for Technology, Policy and Industrial Development, 1990.

Ives, Jane, ed. *The Export of Hazard: Transnational Corporations and Environmental Control Issues*. Boston: Routledge and Kegan Paul, 1985.

Jaffe, Mark. "The West's Latest Export: Unwanted Waste." *Philadelphia Inquirer*, 6 March 1988, C-1.

Juffer, Jane. "Dump at the Border: U.S. Firms Make a Mexican Wasteland." *The Progressive* 52 (October 1988): 24–9.

Lambrecht, Bill. "Trashing the Earth." *St. Louis Post-Dispatch*. Open-ended series; first article, 19 March 1989.

Leonard, Jeffrey H. "Hazardous Wastes: The Crisis Spreads" *National Development* (Australia), April 1986, 33–44.

Mumme, Stephen P. "Dependency and Interdependence in Hazardous Waste Management Along the U.S.-Mexico Border." *Policy Studies Journal* 14 (September 1985): 160–8.

Neff, Alan. "Not in Their Backyards, Either: A Proposal for a Foreign Environmental Practices Act." *Ecology Law Quarterly* 17 (3): 401 (Fall 1990).

Nordquist, Joan. *Toxic Waste: Regulatory, Health, International Concerns. A bibliography*. Santa Cruz, Ca.: Reference and Research Services, 1988.

Organisation for Economic Co-operation and Development. *Transfrontier Movements of Hazardous Wastes: Legal and Institutional Aspects*. Paris: Organisation for Economic Co-operation and Development [Washington, D.C.: sales agents, OECD Publications and Information Center], 1985.

Porterfield, Andrew. "How Two Brothers Dumped the U.S. Government's Hazardous Chemicals on the Third World." News America/Times of London Syndicate, 13 February 1987.

Porterfield, Andrew. "To Tonga, With Love." *California Business*, December 1987, 68–71.

Porterfield, Andrew and David Weir. "The Export of U.S. Toxic Wastes." *The Nation* 245 (3 October 1987): 325.

Postel, Sandra. *Defusing the Toxics Threat: Controlling Pesticides and Industrial Waste.* Worldwatch Paper 79. Washington, D.C.: Worldwatch Institute, 1987.

Rose, Elizabeth C. "Transboundary Harm: Hazardous Waste Management Problems and Mexico's Maquiladoras." *The International Lawyer* 23 (1): 223–44 (Spring 1989).

Spalding, Heather, ed. *The International Trade in Wastes: A Greenpeace Inventory.* 5th ed. Washington, D.C.: Greenpeace, 1990.

Third World Network. *Toxic Terror: Dumping of Hazardous Wastes in the Third World.* Penang, Malaysia: Third World Network, 1989.

Tolan, Sandy. "The Border Boom: Hope and Heartbreak." *The New York Times Magazine*, 1 July 1990, 16.

United Nations Centre on Transnational Corporations. *Environmental Aspects of the Activities of Transnational Corporations: A Survey.* New York: United Nations, 1985. (ST/CTC/55).

United Nations Environment Programme. "Industrial Hazardous Waste Management (Special Issue)." *Industry and Environment* No. 4, 1983.

United States Congress. House. Committee on Government Operations. Environment, Energy and Natural Resources Subcommittee. *International Export of U.S. Waste: Hearing before a Subcommittee of the Committee on Government Operations, House of Representatives, One Hundredth Congress,*

second session, July 14, 1988. Washington, D.C.: U.S. Government Printing Office, 1989.

United States Congress. House. Committee on Energy and Commerce, Transportation and Hazardous Materials Subcommittee. *Waste Export Control: Hearing before the Subcommittee on Transportation and Hazardous Materials of the Committee on Energy and Commerce, House of Representatives, One Hundred First Congress, first session, on H.R. 2525, July 27, 1989*. Washington, D.C.: U.S. Government Printing Office, 1989.

United States General Accounting Office. *Environmental Protection: Bibliography of GAO Documents January 1985-August 1988*. Washington, D.C.: General Accounting Office, 1989. (GAO/RCED-89-23).

United States General Accounting Office. *Hazardous Waste: EPA Has Made Limited Progress in Determining the Wastes to be Regulated. Report to the Chairman, Subcommittee on Commerce, Transportation and Tourism, Committee on Energy and Commerce, House of Representatives*. Washington, D.C.: General Accounting Office, 1986. (GAO/RCED-87-27).

United States General Accounting Office. *Illegal Disposal of Hazardous Waste: Difficult to Detect or Deter. To the Chairman, Subcommittee on Investigations and Oversight, Committee on Public Works and Transportation, House of Representatives/by the Comptroller General of the United States*. Washington, D.C.: General Accounting Office, 1985. (GAO/RCED-85-2).

Victor, Jean Andre. *Sur la Piste des Déchets Toxiques*. [On the Trail of Toxic Wastes]. Haiti: L'Imprimeur II, 1989.

Weir, David. *The Bhopal Syndrome: Pesticides, Environment and Health*. Center for Investigative Reporting. San Francisco: Sierra Club Books, 1987.

Weir, David and Mark Schapiro, *Circle of Poison*. Center for

Investigative Reporting. San Francisco: Institute for Food and Development Policy: San Francisco, 1981.

World Resources Institute. *World Resources 1990–1991.* 5th ed. New York: Oxford University Press, 1990.

Hazardous Waste: General

Arbuckle, J. Gordon et al. *Environmental Law Handbook.* 10th ed. Rockville, M.D.: Government Institutes, 1989.

Blumberg, Louis and Robert Gottlieb. *War on Waste: Can America Win its Battle With Garbage?* Washington, D.C.: Island Press, 1989.

Brown, Michael H. *Laying Waste: The Poisoning of America by Toxic Chemicals.* New York: Pocket Books, 1981.

Brown, Michael H. *The Toxic Cloud.* New York: Harper & Row, 1987.

Bullard, Robert D. *Dumping in Dixie: Race, Class and Environmental Quality.* Boulder, Co.: Westview Press, 1990.

California. Hazardous Waste Management Council. Chairman, Assemblywoman Sally Tanner. *Hazardous Waste Management Plan.* Sacramento: Hazardous Waste Management Council, 1984.

Cohen, Gary and John O'Connor, eds. *Citizen's Toxic Protection Manual.* Boston: National Toxics Campaign, 1988.

Cohen, Gary and John O'Connor, eds. *Fighting Toxics: A Manual for Protecting Your Family, Community and Workplace.* National Toxics Campaign. Washington, D.C.: Island Press, 1990.

Coyle, Dana et al. *Deadly Defense. Military Radioactive Landfills: A Citizen Guide.* New York: Radioactive Waste Campaign, 1988.

EarthWorks Group. *50 Simple Things You Can Do to Save the Earth.* Berkeley, Ca.: EarthWorks Press, 1989.

Epstein, Samuel S. et al. *Hazardous Waste in America.* San Francisco: Sierra Club Books, 1982.

Goldman, Benjamin A. et al. *Hazardous Waste Management: Reducing the Risk.* Council on Economic Priorities. Washington, D.C.: Island Press, 1986.

Jorgensen, Eric P., ed. *The Poisoned Well: New Strategies for Groundwater Protection.* Sierra Club Legal Defense Fund. Washington, D.C.: Island Press, 1989.

Moore, Andrew Owens. *Making Polluters Pay: A Citizen's Guide to Legal Action and Organizing.* Washington, D.C.: Environmental Action Foundation, 1987.

Newsday. *Rush to Burn: Solving America's Garbage Crisis?* Washington, D.C.: Island Press, 1989.

Oldenburg, Kirsten U. and Joel S. Hirschhorn. "Waste Reduction: From Policy to Commitment." *Hazardous Waste & Hazardous Materials* 4 (1): 1–8 (1987).

Phillips, Amanda M. and Ellen K. Silvergeld. "Editorial: Health Effects Studies of Exposure From Hazardous Waste Sites—Where Are We Today?" *American Journal of Industrial Medicine* 8:1–7 (1985).

Piasecki, Bruce, ed. *Beyond Dumping: New Strategies for Controlling Toxic Contamination.* Waterport, CT: Quorum Books, 1984.

Regenstein, Lewis. *America the Poisoned.* Washington, D.C.: Acropolis Books, 1982.

Sarokin, David J., et al. *Cutting Chemical Wastes.* New York: INFORM, 1985.

United States Congress. Office of Technology Assessment. *Facing America's Trash: What Next for Municipal Solid Waste?* Washington, D.C.: U.S. Government Printing Office, 1989. (OTA-0-424).

United States Congress. Office of Technology Assessment. *From Pollution to Prevention; A Progress Report on Waste Reduction.* Washington, D.C.: U.S. Government Printing Office, 1987. (OTA-ITE-347).

United States Congress. Office of Technology Assessment. *Serious Reduction of Hazardous Waste: For Pollution Preven-*

tion and Industrial Efficiency. Washington, D.C.: U.S. Government Printing Office, 1986. (OTA-ITE-317).

United States Environmental Protection Agency. "Introduction to Books in the Hazardous Waste Collection" (report generated from the in-house hazardous waste database). Washington, D.C.: EPA Headquarters Library, 1986.

United States Environmental Protection Agency. Office of Pesticides and Toxic Substances. *Federal Activities in Toxic Substances.* Washington, D.C.: Office of Pesticides and Toxic Substances, 1980. (EPA 560/13-80-015).

United States Environmental Protection Agency. Office of Solid Waste and Emergency Response. RCRA Information Center. *A Catalog of Hazardous and Solid Waste Publications.* 3d ed. Washington, D.C.: Office of Solid Waste and Emergency Response, 1989. (EPA/530-SW-89-054).

United States Environmental Protection Agency. Office of Solid Waste and Emergency Response. *Report to Congress on the Minimization of Hazardous Wastes.* Washington, D.C.: Office of Solid Waste and Emergency Response, 1986. (EPA/530-SW-86-033).

United States Environmental Protection Agency. Office of the Inspector General. *Follow-up on EPA's Program to Control Export of Hazardous Waste.* Washington, D.C.: Office of the Inspector General, 1990. (EPA Audit Report E1DSGO-05-5003-0400011).

International Treaties and Agreements

"Agreement Between the Government of the United States of America and the Government of Canada Concerning the Transboundary Movement of Hazardous Wastes," Ottawa, 28 October 1986. (U.S. Department of State, Office of Treaty Affairs).

"Agreement Between the United States of America and the United Mexican States on Cooperation for the Protection and Improvement of the Environment in the Border Area," La Paz, 14 August 1983. (Treaties and Other International Acts Series 10827). See Annex III, "Agreement of Cooperation Between the United States of America and the United Mexican States Regarding the Transboundary Shipments of Hazardous Wastes and Hazardous Substances," Washington, D.C., 12 November 1986. (U.S. Department of State, Office of Treaty Affairs).

"Basel Convention on the Control of Transboundary Movements of Hazardous Waste and their Disposal," Switzerland, 22 March 1989. United Nations Environment Programme. (UNEP/IG.80/3).

"Decision and Recommendation of the Council on Transfrontier Movements of Hazardous Waste," Organisation for Economic Co-operation and Development Council, Paris, 1 February 1984. (OECD Doc. C [83] 180).

"Decision-Recommendation of the Council on Exports of Hazardous Wastes from the OECD Area," Organisation for Economic Co-operation and Development Council, Paris, 5 June 1986. (OECD Doc. C [86] 64).

"Directive on the Supervision and Control Within the European Community of the Transfrontier Shipment of Hazardous Waste," Council of the European Economic Community, 13 December 1984. (*Official Journal of the European Communities* No. L 326).

"The Fourth ACP (African, Caribbean and Pacific States–EEC (European Economic Community) Convention" ("Lome IV Convention"), Lome, Togo, 15 December 1989. (*The Courier* [European Community] No. 120, March/April 1990). See "Joint Declaration on Article 39 on Movements of Hazardous Waste or Radioactive Waste."

Documentaries

"Dispatches: The Toxic Trail" (focus on waste dumping in Africa, particularly Benin). 30 minutes. Producer/Director John Longley. London: Box Productions for Channel 4, 1989.

"4 What It's Worth: Toxic Waste" (the export of Australian waste to Rechem's treatment plant in Wales). 30 minutes. Producer/Director Mark Edmondson. London: Thames Television in association with the Australian Broadcasting Corporation, 1989.

"How Green is the Valley" (mercury exports to South Africa). 30 minutes. Washington, D.C.: Greenpeace, 1990.

"Inside the Poison Trade." 45 minutes. Producer Philip Brooks. London: Central Independent Television in association with Television Trust for the Environment (TVE), 1989.

"Marina di Carrara" (efforts to export hazardous wastes from Italy backfire). 13 minutes. Hamburg: Spiegel-TV, 22 June 1989.

"Veleni D'Oro" ["Golden Poisons"](attempts to export hazardous wastes from Italy). 30 minutes. Producer Lamberto Sposini. Rome: RAI Corp., 1989.

Organizations

Chemical Manufacturers Association
2501 M Street NW
Washington, DC 20037
202/887-1100

A nonprofit trade association with 200 members that administers research and conducts studies, workshops, and technical symposia. Operates Chemical Transportation

Emergency Center (CHEMTREC) and has an International Trade Committee.

Citizens Clearinghouse for Hazardous Wastes
PO Box 926
Arlington, VA 22216
703/276-7070

Helps citizens' organizations around the world develop strategies for building grassroots movements and international links.

Citizens for a Better Environment
33 East Congress, Suite 523
Chicago, IL 60605
312/939-1530

A national environmental organization that has done extensive research on hazardous waste, spills, and dumpsites.

Environmental Defense Fund
1616 P St., NW, Suite 150
Washington, DC 20036
202/387-3500

A national environmental advocacy organization with an interest in solid-waste management. Focuses primarily on environmental policy at the federal level.

Environmental Research Foundation
PO Box 3541
Princeton, NJ 08543
609/683-0187

Privately funded information center that keeps a computer database called RACHEL (Remote Access Chemical Hazards Electronic Library) of information on hazardous waste and publishes *Hazardous Waste News* ("Weekly news and resources for citizens fighting toxics").

Greenpeace International Waste Trade Project
1436 U Street, NW
Washington, DC 20009
202/462-1177

Greenpeace, a global environmental organization with over 2.5 million members, is seeking a global ban on the trade of all wastes that will release pollutants into the environment. At the same time, it supports and encourages bans of waste import and export in individual countries. Publishes the *Greenpeace Waste Trade Update* ("A quarterly newsletter on the international waste trade").

International Maritime Bureau
Maritime House, 1 Linton Road
Barking, Essex 1G11 8HG
United Kingdom
44-01-591-3000

A division of the International Chamber of Commerce, the IMB is a nonprofit membership organization that fights fraud, violence, and malpractice in the global shipping and transportation industry. IMB operates a Waste Hotline to gather information on shipments of hazardous chemicals and the unregulated dumping of hazardous wastes at sea, and publishes a monthly report, the *ICC Commercial Crime International*.

International Organization of Consumers Unions
PO Box 1045
Penang, Malaysia
60-4-20391

An international network of consumers' and citizens' groups concerned with hazardous products and substances. Publishes newsletter *Consumer Currents* as well as periodic reports on specific topics including the hazardous-waste trade.

National Toxics Campaign
1168 Commonwealth Avenue, 3rd Floor
Boston, MA 02134
617/232-0327

An advocacy organization that promotes cutting waste production and protecting public health by providing technical and organizing assistance and environmental testing. Publishes quarterly newsletter *Toxic Times*.

Natural Resources Defense Council
40 West 20th Street
New York, NY 10011
212/727-2700

A national environmental group that monitors and promotes federal environmental legislation, EPA's enforcement of environmental laws, and international toxic-export treaties and agreements.

Organization of African Unity
PO Box 3243
Addis Ababa, Ethiopia
251-1-517700

346 East 50th Street
New York, NY 10022
212/319-5490

A coalition of African nations that has spoken out against international toxic-waste dumping.

United Nations Environment Programme (UNEP)
Head Office
PO Box 30552
Nairobi, Kenya
254-2-333930

North American Office
United Nations
Room DC2-0816
New York, NY 10017
212/963-8138

An agency of the United Nations that coordinated the Basel Convention and monitors international environmental issues.

About the Authors

THE CENTER FOR INVESTIGATIVE REPORTING, founded in 1977, is the only independent, non-profit organization in the United States dedicated to doing investigative reporting. From offices in San Francisco and Washington, D.C., CIR staff write for leading newspapers and magazines, and work with television news programs in the United States and abroad. Center reports have helped spark Congressional hearings and legislation, U.N. resolutions, public interest lawsuits, and changes in the activities of corporations and government agencies. Its stories have received dozens of journalism awards, including the National Magazine Award, the National Press Club Award, and the Investigative Reporters and Editors Award.

Among the stories CIR has reported over the last decade are: the export of banned and restricted pesticides to the Third World; the secret history of Soviet and U.S. nuclear accidents at sea; hidden health hazards for workers in automated offices and the electronics industry; the FBI's renewed surveillance of White House critics; and the involvement in the narcotics trade of key Salvadoran political figures linked to death squads. For more information and a publications list, contact CIR at 530 Howard Street, Second Floor, San Francisco, California 94105.

BILL MOYERS, one of America's veteran journalists, has concentrated on the social, political, and international issues facing the United States during his nineteen years in broadcasting. He has been executive editor of the acclaimed series, *Bill Moyers' Journal*; senior news analyst for the CBS Evening News; and chief correspondent for the highly-regarded documentary series *CBS Reports*. His public television series, *A Walk Through the 20th Century*, was named the Outstanding Informational Series of the year (1984) by the National Academy of Television Arts and Sciences.

Since Moyers' return to public television in the fall of 1986, among the programs his Public Affairs Television, Inc. has produced are: *In Search of the Constitution, The Secret Government . . . The Constitution in Crisis, Joseph Campbell and the Power of Myth, The Public Mind, From D-Day to the Rhine*, and *A World of Ideas*. Deputy director of the Peace Corps under President Kennedy, Moyers also served President Johnson as special assistant and then as press secretary. On leaving politics, he became publisher of *Newsday*, which won thirty-three major journalism awards, including two Pulitzers, during his tenure.

GLOBAL DUMPING GROUND is the first independent documentary production of the Center for Investigative Reporting. The program is part of the PBS series FRONTLINE and is a co-production of CIR and public television station KQED in San Francisco. Bill Moyers is the executive editor and correspondent. Videocassette copies of *Global Dumping Ground* are available for $29.95 each, plus $3.00 shipping and handling (California residents please add $2.17 tax), from CIR Video Sales, 530 Howard Street, Second Floor, San Francisco, California 94105.

Index